AFFORESTATION IN INDIA

AFFORESTATION IN INDIA

DIMENSIONS OF EVALUATION

H S Gupta • V K Sinha • R K Singh • D K Sharma

teri

The Energy and Resources Institute

ISBN 978-81-7993-463-0

Suggested citation
Gupta, H. S., V. K. Sinha, R. K. Singh, and D. K. Sharma. 2013. *Afforestation in India: dimensions of evaluation.* New Delhi: TERI

Published by

The Energy and Resources Institute (TERI)	***Tel.***	2468 2100 or 4150 4900
TERI Press	***Fax***	2468 2144 or 2468 2145
Darbari Seth Block		India +91 • Delhi (0) 11
IHC Complex, Lodhi Road	***Email***	teripress@teri.res.in
New Delhi – 110 003	***Website***	www.teriin.org
India		

Printed in India

Foreword

The National Afforestation Programme (NAP) is one of the most ambitious attempts ever made by the Government of India for greening the country with an approach that is local, participatory, transparent, and attempts widespread decentralization. The efforts under this programme have already started yielding results. But it is high time to find out to what extent the NAP objectives are being met and assess its impact on local villagers and green resources developed. The report of independent evaluators and the findings of concurrent evaluation are critical in this respect and, hence, are emphasized in NAP.

Although many generic issues related to the monitoring and evaluation of this programme are normally being addressed by the evaluators according to the NAP guidelines, a better approach would be to adopt a holistic framework for evaluations. Further, the second-order effects, which were not anticipated at the project formulation stage, need to be studied, assessed, and given serious thought as they can affect the overall outcome.

It is also important that the monitoring and evaluation time frame should be designed at the beginning of each project and followed rigorously. The second-order effects should also be suitably incorporated in the evaluation framework. Hence, it is high time that the ongoing monitoring and evaluation process aims at capitalizing the NAP's strengths, opportunities, and threats (keeping in mind the weaknesses) so that it helps the implementers in achieving the desired objectives. The learning from such concurrent, mid-term, and terminal monitoring and evaluation report must be used in formulating and implementing similar future projects in a better manner.

In this background, the book has distilled generic lessons from a meta-assessment of monitoring and evaluation related issues of NAP done so far. Its findings will be of great use to implementers and more to policymakers in formulating greening schemes in the country. It is hoped that newer evaluation methodologies would be capable of addressing the needs and requirement of practitioners, evaluators, and other stakeholders. It is expected that this analytical study would prove very useful to all the stakeholders associated with the forest development agencies in implementing not only the NAP but other such projects as well.

Dr R B Lal

Director, Indian Institute of Forest Management, Bhopal

Foreword

The National Afforestation and Eco development Board (NAEB) has always strived for the highest standard of afforestation work, which also ensures livelihood opportunities to the weaker section of societies living in and around forests and other inaccessible villages. For this reason, it has naturally given high priority to an effective but easy-to-implement monitoring and evaluation mechanism, which ensures high quality output. Since the beginning, the NAEB has followed the established methodology to conduct evaluation in respect of ongoing and completed projects under its different schemes through its pool of evaluators, comprising reputed private organizations, autonomous organizations of the government, and experienced individuals.

In order to execute any plan of work, evaluation/monitoring activity is essential to bring it to the desired end. But "evaluation" is perhaps the most misunderstood word in any programme implementation. Many implementers are suspicious of evaluation because they see this aspect as something that is imposed on them and that acts as an undesired fetter to their freedom. The term "evaluation" has also taken the form of a quantitative concept as it is largely considered to be an important tool of management.

Evaluation is diacritical in any programme involving public finance, including afforestation. Accountability to the Parliament makes it imperative to provide information to policymakers so that the programme and practices are adopted to meet the continuously evolving needs of the society and people. This increases the importance of the monitoring and evaluation aspect of afforestation activities.

The Ministry of Environment and Forests (MoEF) earlier launched a pilot scheme called Samanvit Gram Vanikaran Samriddhi Yojana (SGVSY) in the last phase of the Ninth Five-year Plan, putting great faith in the principles of people's participation. The SGVSY was implemented at a later stage through the newer and decentralized mechanism of the forest development agencies (FDAs), with the basic aim of integrating the afforestation and eco-development activities of various ongoing schemes of the MoEF. The initial experience of this innovative experimentation in the pilot phase had given insight to many deficiencies in the earlier approach followed till the Ninth Five-year Plan. Thus it has helped the MoEF in launching the SGVSY in the Tenth Five-year Plan at a larger scale, by subsuming the other centrally sponsored afforestation schemes of the MoEF. The new version of SGVSY, renamed the National Afforestation Programme (NAP) in the Tenth Five-year Plan, is being implemented all over the country through the two-tier decentralized structure of the FDAs at the forest division level and the village forest committees (VFCs)/joint forest management committees (JFMCs) at the village level in various states and union territories of the country. With the growing spread of the NAP, it becomes pertinent to assess the efficacy of the programme. As part of the monitoring and evaluation exercise, the NAEB has conducted numerous concurrent and independent evaluations of many FDA projects throughout the country.

It is very important that all the evaluation findings and knowledge gained from the implementation of NAP/SGVSY so far should be discussed. With this objective, various interactive workshops were held at various locations in India in 2005/06, which were attended by chairpersons/chief executive officers of the concerned FDAs, senior officers from state forest departments, including forest secretaries, principal chief conservator of forests, and nodal officers in charge of NAP in states. This was further followed up at many newer locations at the regional level in 2006/07, where the independent evaluators engaged in the evaluation of different FDA projects were invited to make presentations of their findings in the field, take part in discussions, and report appropriate follow-up action taken by them. There were a series of seven such workshops coordinated by seven regional centres of the NAEB, based at Solan, New Delhi, Bhopal, Shillong, Kolkata, Mumbai, and Bangalore.

The present book is an attempt to present summarized analyses of the deliberations taken place at all the seven regional workshops conducted from July to October 2005. These interactive forums were unique as they provided an opportunity to various stakeholders involved in NAP to deliberate freely for suggestions to improve the outcome of NAP. It is hoped that the current book will reach all these stakeholders and will serve as a new benchmark for all future evaluation of FDAs and will also stimulate mid-course review and corrections of the NAP.

AR Chaddha
Inspector General of Forests,
Ministry of Environment and Forests,
Government of India, New Delhi

Preface and Acknowledgements

People all over the world are appreciating the afforestation efforts being carried out in India. Agencies such as the Food and Agricultural Organization have acknowledged these efforts and have ranked India in the highest bracket. Although India has a very rich cultural and historical tradition of growing trees and conserving forests, afforestation efforts got a boost in the era of modern scientific forestry. This progress and the current stage of the technology-and-management aspect of afforestation in India are the result of sustained efforts spanning more than 150 years of scientific forestry.

There has been commendable progress in the field of afforestation after India's independence. Attempts have been made to match the supply and demand of products derived from forests, according to the aspirations of developing India. This path of afforestation has been fraught with many challenges such as financial crunch, climatic variation, edaphic variability, and biotic interferences. But the dedicated efforts of the forest departments of the centre and states, forestry research institutions, forestry-oriented non-governmental organizations, and the civil society have contributed to the current body of knowledge and practice. We should, particularly, appreciate the critical role of "monitoring and evaluation" activities in the entire process of afforestation. The findings of the "monitoring and evaluation" exercises have always helped and would be important in the future for sustainable man-made forestry.

This book draws on the efforts made by numerous practitioners and planners involved in the National Afforestation Project, who have planned, formulated, implemented, supervised, monitored, evaluated, or associated in different capacities with this ambitious project of the Ministry of Environment and Forests (MoEF), Government of India. We appreciate and thank the vital inputs provided by all these people in different regional interactive workshops held during 2005 and 2006, necessarily in an impersonal manner.

We very much appreciate the vital inputs provided at critical stages by the principal chief conservator of forests of various state forest departments in India in general and the nodal officers (of forest development agency-related activities) Shri A S Joshi, Shri R N Saxena, and Shri R P Singh of the Madhya Pradesh Forest Department; Shri B P Singh of the Orissa Forest Department; Dr A Bhalla and Shri R K Tamta of Chhattisgarh State Forest Department in particular. They had not only participated and provided key ideas in different workshops but also ensured the wider participation of key functionaries associated with the NAP in these workshops to articulate their free and frank views on various facets of the NAP.

We also thank Shri A R Chaddha, Inspector General (IG), National Afforestation and Eco-development Board (NAEB); Shri Ashok Pai, former Deputy Inspector General (DIG), NAEB; and Dr Sanjay Kumar, former DIG, MoEF, Government of India, for reviewing the draft of this document, offering valuable comments from time to time, and providing their kind guidance. All of them have been key motivators for the editorial team for coming out with such an analytical work, and we sincerely acknowledge it.

The book has seen the light of the day through the financial and moral support of the NAEB, MoEF. Moreover, the secretary, MoEF; director general (DG, forest), Government of India; IG, NAEB; and other related officers of the MoEF deserve sincere thanks for their patronage and support. We are also thankful to the Director, Indian Institute of Forest Management (IIFM), Bhopal, for encouraging and facilitating us in this endeavour.

The book could not have been brought out without the constant guidance and support provided by P J Dilipkumar, DG (forest), Government of India, and K B Thampi, former IG, NAEB, at every stage of its evolution.

The opinions reflected in this book are those of the authors and not necessarily those of the NAEB, MoEF, or IIFM.

Editors
H S Gupta
V K Sinha
R K Singh
D K Sharma

Disclaimer

Contents

List of Tables

List of Figures

List of Boxes

List of Abbreviations

ANR	assisted natural regeneration
BPL	below poverty line
CBO	community-based organization
CCF	chief conservator of forest
CEO	chief executive officer
CF	conservator of forest
CIG	community interest group
CPM	critical path method
DFO	divisional forest officer
ECS	electronic clearing system
EDC	eco-development committee
EOP	end of project
EPA	entry-point activities
FAO	Food and Agriculture Organization
FD	forest department
FDA	forest development agency
FSI	Forest Survey of India
FY	financial year
GoI	Government of India
GoO	Government of Odisha
GPS	geographical positioning system
GTZ	Deutsche Gesellschaft für Technische Zusammenarbeit (German Development Cooperation)
IFAD	International Fund for Agriculture and Development
IIFM	Indian Institute of Forest Management
IVI	importance value index
JFM	joint forest management

JFMC	joint forest management committee
KBK	Kalahandi, Bolangir, Koraput
LOG FRAME	logical framework
M&E	monitoring and evaluation
MoEF	Ministry of Environment and Forests
MoRD	Ministry of Rural Development
MOV	means of verification
MTR	mid-term review
MWS	micro-watershed
NA	not available
NAEB	National Afforestation and Eco-development Board
NAP	National Afforestation Programme
NGO	non-governmental organization
NTFP	non-timber forest product
OBC	other backward class
OVI	objectively verifiable indicators
OWDM	Orissa Watershed Development Mission
PCCF	principal chief conservator of forests
PERT	programme evaluation and review technique
PIA	project implementing agency
PRA	participatory rural appraisal
PRI	Panchayati raj institutions
PY	project year
RRA	rapid rural appraisal
SC	scheduled caste
SE	standard error
SFD	state forest department
SFM	sustainable forest management
SGVSY	Samanvit Gram Vanikaran Samriddhi Yojana
SHG	self-help group
SMC	soil and moisture conservation
ST	scheduled tribe
VDA	village development agency
VDF	village development fund

VFC	village forest committee
WDC	watershed development committee
WORLP	Western Orissa Rural Livelihoods Project
WSM	watershed management

CHAPTER 1

Monitoring and Evaluation Aspects of Forestry Projects: A Review

H S Gupta

INTRODUCTION

The purpose of any developmental project is to bring about positive changes in the present status. In the past, various forestry projects oriented towards community, social, or rural development had been tried, which had a different set of objectives, activities, and management style than traditional afforestation projects. Most of these projects had environmental and protection objectives like traditional forestry projects. Later, some additional objectives were included to increase rural livelihood opportunities and raise the living standards of villages located at the fringe of forests by trying to route income and welfare benefits to the poorest sections among them.

Different project objectives demand different approaches to project management. In a traditional forestry project, decisions are normally made by the top management and carried out by the field functionaries of the forest department. In the new paradigm, whereas, many of the decisions and their execution involve both the management and the implementer, whose views should be sought on all important issues. This calls for a different monitoring and evaluation (M&E) system as well.

It is emphasized that M&E systems need to have a wider base, which perhaps makes projects incorporating these systems more difficult to operate in comparison to a traditional forestry project. However, they are particularly important in effective project implementation. Monitoring and evaluation are important components of a project planning cycle. Figure 1.1 shows the monitoring and evaluation relationship to other planning activities.

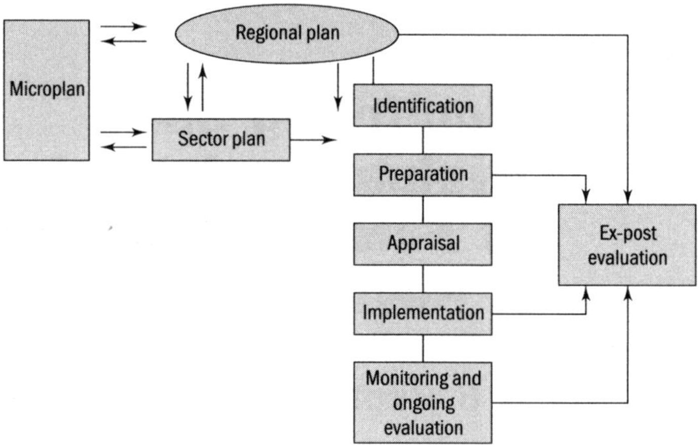

Figure 1.1 *Schematic representation of project planning and operation activities*
Source *Adapted from Clayton and Petry (1984)*

In this chapter, an effort has been made to review and discuss the principles and procedures of M&E, and how they can be related to the National Afforestation Programme (NAP).[1]

NATIONAL AFFORESTATION PROGRAMME

In the last phase of the Ninth Five-year Plan, the Ministry of Environment and Forests (MoEF) launched a pilot scheme called Samanvit Gram Vanikaran Samriddhi Yojana (SGVSY), putting great faith in the principle of people's participation. The SGVSY was implemented at a later stage through the newer and decentralized mechanism of the forest development agencies (FDAs), with the basic aim of integrating afforestation and eco-development activities of various ongoing schemes of the MoEF. The initial experience of this innovative experimentation in the pilot phase had given insight into many deficiencies in the earlier approach followed till the Ninth Five-year Plan. Thus, it has helped the MoEF in launching the SGVSY in the Tenth Five-year Plan at a larger scale, by subsuming all other centrally sponsored afforestation schemes. The new version of SGVSY was renamed as the National Afforestation Programme. It is being implemented nationwide through the two-tier decentralized structure of FDAs, namely at the forest division level and the village forest committees (VFCs)/joint forest management committees (JFMCs) at the village level.

[1] For details of NAP, refer to Appendix 2.

The NAP is not only for catering to the basic needs of communities for fuelwood, fodder, and other wood and secondary forest products through the FDAs, but also for better management of forests to increase production and protect the environment and soil. Conventionally, all the active participants and intended beneficiaries of forestry projects, which comprise villagers, state forest departments (SFDs), VFCs, and implementing agencies (here FDAs), are also monitored on the basis of basic principles, which have been used in other forestry projects. But given the innovative approach of NAP, the M&E aspect of it has to be devised to capture these innovations.

ELEMENTS OF M&E IN FORESTRY ACTIVITIES

Conventionally, project M&E activities have been designed to help the MoEF and SFDs in charting the progress of project implementation, to achieve the intended outputs according to the programme objectives, and to take corrective action when implementation deficiencies are detected by the M&E process. It has, thus, enabled the MoEF and SFDs to assess the relevance, efficiency, and effectiveness of a programme, together with its impact on villagers and the environment. Monitoring and evaluation can also provide guidance and examples to plan the future course of any programme, including NAP.

Some dimensions of M&E of forestry activities are reflected in the work of the International Fund for Agricultural Development (IFAD) (1984). Monitoring is a continuous or periodic surveillance of the implementation of a project to ensure that input deliveries, work schedules, targeted outputs, and other required actions proceed according to the plan (IFAD 1984).

The purpose of M&E is to assist field managers in achieving efficient and effective scheme operation and performance, by providing them with appropriate information and feedback in relation to the critical activities of a scheme. Thus it provides the management a basis for taking timely corrective action by identifying constraints and inadequacies in the performance of the project. Monitoring can be considered a part of the management information system or an internal component of the management decision-making.

Further, evaluation is a systematic process that attempts to assess, as objectively as possible, the relevance, effectiveness, and impact of a project in the context of the project objectives (IFAD 1984). To elaborate further, evaluation essentially analyses the rationale and logic of the project (including its objectives/design), reviews the implementation process (inputs, effects, and impact), and assesses, in the light of the foregoing, the validity and relevance of project objectives/design and the project effectiveness and efficiency in achieving the intended results.

As the aforementioned approach suggests, evaluation is a learning and action-oriented management tool that seeks to improve the effectiveness, relevance, and impact of currently operating projects. It also provides information (in concert with monitoring) for decision-makers to make any needed adjustments to the objectives, activities, operation, and performance of an ongoing project. It includes examination of assumptions to check whether they are valid, because the planners might have got them wrong, unforeseen factors might have made them invalid, or experience with the project may require their redefinition.

It is evident that monitoring and evaluation are closely linked and they provide reliable, relevant, and timely information for decision-making and solving specific problems, which are of importance and concern to the management and participants of a project. This necessitates that M&E systems are flexible in their use for different types of projects. Hence, the design of M&E systems must continuously change to perform their problem-solving role.

Ideally, a complete M&E system should extend throughout the project and the administrative hierarchy, from the MoEF to the field level. At higher levels, it can be more "strategic" and concerned with issues such as overall progress, major problems, budget disbursements, and the planning of future projects. As the M&E information passes up the hierarchy, it can be increasingly condensed for strategic surveillance purposes through periodic reports.

Experience indicates that with M&E, the value of information systems becomes increasingly clear and important. It is an important tool employed for improving project effectiveness. The spread of M&E systems can increasingly expose the flimsy base upon which much project planning rests. However, on the positive side, the exposure of these defects is likely to prove an important feedback for improving the planning of projects in the future.

The operation of an M&E system can bring in additional benefits to a project as a spin-off. It ensures improved work planning, better awareness among stakeholders, incentive for improved staff performance, and improved project "image".

BASIC ELEMENTS OF M&E IN FORESTRY OR AFFORESTATION PROJECTS

An M&E system normally focuses on five elements of equal importance: operation, performance, effect, impact, and context of a project. Project operation includes the tasks and activities performed regularly or intermittently that are essential for the prescribed functioning and implementation of a project. For example, it may include the delivery and distribution of project inputs such as fertilizers and seedlings, activities such as planting and hoeing, financial flows, and staffing.

A forestry project performance is measured by the outputs resulting from the project operation. It may include such aspects as nurseries constructed, seedlings produced and distributed, area planted with trees, forest area managed, and the number of checkdams/stopdams created.

Project effects are the outcomes of project operation and performance, and include immediate project objectives and goals. Effects can include more trees grown, increased supply of fuelwood, labour time saved in collecting forest products, providing the forest product for the needs of villages at the fringe of forests, and so on.

Project impact is the result or consequence of project performance, performance, and effects. Impact relates the results of a project to its long-range objectives or goals, and indicates the extent to which they have been achieved. It denotes changes in the status of beneficiaries resulting from a project—for example, increase in family income, contribution of forest products in nutrition, and change in living standards. It may include the achievement of larger welfare objectives such as increased literacy and wider participation of project beneficiaries in project decision-making. Project impact is further affected by unplanned changes in the local environmental and economy that result from implementation of the project. For example, soil erosion, environmental damage to wildlife and natural flora, forest resources, and catchment areas can have adverse effects on the price of forest inputs and products.

Monitoring and evaluation of the impact of a project require a longer time horizon than other elements. Some impact changes may be detected during the implementation of the project, for example, increased income from medicinal plants or other non-timber forest products (NTFPs) may be assessed quickly on a very successful scheme. But in other cases, the full impact of a project cannot be assessed in a substantial way until some years after its full development or completion have been achieved. Forestry projects normally have long time horizons because of the relatively long period required for many trees to reach maturity.

Project context relates to the physical and socio-economic "situation" to which the project is intended to respond, the attitudes of the rural groups to the activities and objectives of a project, and the activities of the project and non-project people, which are relevant to the project objectives. This element of M&E attempts to test the validity or relevance of project objectives and their related activities.

The focus in the case of a traditional forestry project was normally on monitoring project inputs, outputs, and financial flows, with limited attention given to the evaluation of project context and impact. In contrast, in the NAP there will be more emphasis on the evaluation of project effects and context, besides conventionally monitoring project inputs and activities. The NAP also intends to increase the involvement of rural people through participatory microplanning and implementation efforts. This basically involves discussing with the participants and other stakeholders their views of the assumptions, objectives, and effects of the project.

The monitoring and evaluation of project operation, performance, effects, and impact are mainly based on the observation and verification of variables or indicators of project inputs, outputs, activities, and effects as well as of project objectives, external factors, and constraints. Hence, their identification and measurement are very critical.

The monitoring and evaluation of "critical external factors" also cannot be neglected. These factors are exogenous to a project and can have unexpected effects on it. Market prices of purchased inputs, project outputs, and weather conditions are examples of these. Other external factors that can adversely affect project operation and performance, and are outside the control of management, include domestic inflation, shortages and late delivery of materials, failure to provide infrastructure (such as electricity or a road connection to a nursery site) to a project on time, distortion of forestry price policies, and other market aberrations.

Sometimes it is not possible to directly observe and measure project monitoring indicators. There are many reasons for this. Some of them include the high cost or practical difficulties of collecting information (field surveys, for example, can be costly); difficulty in distinguishing the effects of a project input or activity; and impracticability of direct measurement in the case of a long-time horizon of timber production. Similarly, income surveys of rural households are often costly to undertake and rather slow to produce results.

The yield and production of most forest products are linked to the long period it takes for trees to grow. It is, therefore, not possible in the short and medium term to use the direct effects and impact indicators of forestry projects; instead, indirect or proxy indicators for these are used, such as seedling survival rates and tree growth rates.

The process of identifying and selecting monitoring indicators has an important influence on the usefulness of the M&E system. It also affects the efficiency and effectiveness of project implementation, because it determines the total amount of information in the system. The full range of indicators derived from the project objective structure and consideration of unexpected effects will often far exceed the capacity of the M&E system to handle them. Indicator selection, therefore, can be influenced by constraints such as extreme difficulty in the collection and measurement of information, motivation and ability of staff affecting the accuracy of information collected, and so on. At the same time, the demands of the major users of the system must receive sufficient consideration. Data requirements thus become a function not only of what should logically be collected and measured, but also of relevance, measurability, feasibility, timeliness, and simplicity (Bhattarai and Campbell 1984).

Some ecological, social, and economic indicators related to forestry projects have been listed in Box 1.1 (Bahuguna and Upadhyay 2004). They have become very relevant both in the context of principles of sustainable forest management and to the NAP.

DESIGN AND IMPLEMENTATION OF M&E SYSTEMS FOR FORESTRY/ AFFORESTATION PROJECTS

Forestry projects vary in size, organization, and purpose. But, in general, the common aims can be listed as follows.

- To improve the living standards of villagers.
- To promote the self-reliance of rural people by their active participation in the management of forest resources.
- To avoid or reduce environmental degradation and conserve soil and water resources.

The pursuit of these objectives usually involves the following types of project activities.

- Construction and operation of nurseries for seedling production and distribution.
- Establishment of different types of plantations.
- Distribution of seed or seedlings to rural people for their own planting.
- Construction of different types of structure for soil and moisture conservation.

BOX 1.1 | Monitoring and evaluation: relevance of some parameters in forestry projects

Indicators to Monitor

The parameters that need to be monitored during the implementation of the programme can be classified as directly measurable and incidental.

Directly Measurable Parameters

Rehabilitation of forest

The amount of degraded forests rehabilitated and converted into forests with good site quality would be a good indicator of the success of the programme. The indicators to monitor include the following.

- Nature of regeneration status.
- Change in floral composition.
- Changes in soil fauna and wildlife (including biodiversity indices).
- Importance of value index of vegetation.
- Change in canopy cover.
- Number of trees per hectare.
- Basal area of tree crop per hectare.
- Top height of trees.
- Average crop height.
- Timber volume per hectare.

Similar changes in the composition of vegetation, status of regeneration, and further improvement in the site quality of good forests (crown density of 40% and above) should be monitored regularly, especially with reference to sustainability of non-timber forest products (NTFPs).

Changes in species composition

Changes in species composition include the natural stage of succession in which the forest is and the changes induced by protection and choice of appropriate species suited to the climate and soil of the particular area as well as to the needs of the local people.

Status of biodiversity

The extent of biodiversity reflects the condition of a forest and the measures taken for its protection. Extensive mapping of the biodiversity of the forests in the country and its periodic revision would provide a good indicator for monitoring.

Change in socio-economic status of the people

Change in the socio-economic status of people is an important indicator from the point of view of people and sustainability of the programme. The changes

Contd...

BOX 1.1 | ***Contd...***

in employment, income and migration levels, occupation and land use patterns, health of humans and cattle, introduction of stall feeding of cattle, level of firewood and NTFP collection for self-use and employment, introduction of alternative sources of energy, marketing of forest produce, status of women and other weaker sections, and social harmony should be regularly monitored and documented.

Incidents of forest fires and illicit removal of forest produce

Forest protection is one of the major objectives of any project. Reduction in incidents of forest fire and illicit felling would be a good indicator for the success of the programme.

Functioning of joint forest management committees

Constant and extensive monitoring is called for to ascertain whether the joint forest management committees (JFMCs) are functioning in a democratic and transparent manner. Some indicators are as follows.

- Number of JFMC members as a proportion of the total village population.
- Proportion of members from the weaker sections such as women, scheduled castes, scheduled tribes, and landless people.
- Percentage of elected members (as against nominated, ex officio members) in the executive committee.
- Percentage of members from the disadvantaged sections in the executive committee.
- Participation of women.
- Frequency of meetings and visits made by senior forest officers.
- Actual attendance in these meetings, especially of the disadvantaged sections.
- System of decision-making.
- Role of local political parties and their leaders.
- Maintenance of accounts and records of money and other assets.
- Internal/external audit of accounts.
- Awareness of JFMC members about the financial, management, and organizational aspects of the committee.
- High points or decisions made in the meeting.
- Implementation of the decisions made in the meeting.
- Networking with panchayats and other government departments.
- Synergy in action and approach of working.

Contd...

BOX 1.1 | ***Contd...***

Development of JFMC into strong and self-sustaining institutions

The ultimate goal of the joint forest management programme is to help the people management committees attain the strength to become self-sustaining institutions. Some indicators are as follows.

- Legal recognition of JFMCs.
- Provision of a memorandum of understanding with the forest department.
- Well-defined powers and responsibilities of JFMCs vis-à-vis other agencies.
- Common fund and common assets and their management procedure by JFMCs.
- Level of dependence on the forest department.
- Level of dependence on other organizations such as non-governmental organizations.
- Relations with *Panchayati Raj* institutions and role of local leadership.
- Transparent working.

Incidental Parameters

Incidental parameters include the following ecological indicators.

- Soil profile.
- Groundwater level.
- Impact on the productivity of primary sector activities such as agriculture, animal husbandry, and horticulture.
- Any other environmental parameters affecting vegetation.

Source Adapted from Bahuguna and Upadhyay (2004)

- Enhanced awareness campaign to improve the establishment and maintenance of trees (including increased forest production within plantations), and encouragement to elicit more active participation of villagers in project activities and forest management.

This wide range of objectives and activities imply the need for a comprehensive M&E system based on the type of project. The focus should be on the surveillance and assessment of project inputs, outputs, effects, and context, and to a lesser extent on impact. Close attention has to be paid to the number of nurseries constructed, seedlings distributed, number of hectares planted, and so on. These are the indicators that reflect project inputs and outputs. Some indicators that reflect project effects include the survival of seedlings by species, the number of trees planted and surviving, increased output of forest products, and increased grass and forage production. The effects and context of M&E of a project should also focus on the prevailing forestry practices and attitudes of the people to project objectives and activities.

ASSESSING EFFECTS AND IMPACTS THROUGH M&E

In afforestation projects, some included impacts of M&E are as follows: surveillance and assessment of timber and forest products produced against people's need; increased income and living standards of project participants; increased participation of villagers in management of the project; and improvement in the physical environment and its resources. The studies that measure and assess these effects and impact indicators require significant monitoring resources for baseline and ad hoc field surveys. Hence, they can be undertaken only once or infrequently during project implementation or otherwise specific provision for the same as made in project.

The monitoring and evaluation of project effects, impact, and context can be based on surveys, which are relatively easy to design and execute and can be completed in a relatively short period of time—for example, surveys of seedling and plantations survival rates, the extent of private tree planting, and species preference. Participatory M&E is somewhat different in the sense that it is continuous and based on an informal survey approach. But its advantage is that it can provide rapid feedback on implementation problems and success, as well as on adverse effects and impact.

M&E OF SOCIO-ECONOMIC PARAMETERS

The monitoring and evaluation of socio-economic effects are also important but were often neglected in the past. There is an element of uncertainty in socio-economic effects, which is difficult to predict. Also, unless the project has been operating for some time, there is an uncertainty about what is to be mentioned.

Unplanned socio-economic effects of a project cover issues such as the demand for project inputs and the supply of forest products, and their impact on the respective markets, all of which have their bearing on projects. Occasional market studies are, therefore, required to monitor changing conditions that might follow project implementation. Continuous monitoring of input and output prices is also necessary.

OTHER IMPORTANT ASPECTS OF M&E

An important aspect of the implementation of an M&E system is determination of who should undertake it. It can be carried out by the forest departmental staff or by a separate M&E unit. When the departmental staff undertake M&E activities, their influence at the design stage will normally be expected to keep the system as simple and cheap as possible—to collect only relevant and useful information and deliver it to the right place, in a form that ensures its operational use. Departmental staff often complain that excessive M&E adversely affect their normal work. This problem can be avoided. Project staff are probably less disposed to question project assumptions or evaluate their own performance, which may need attention in the design and implementation stage.

Hence, a separate unit consisting of qualified staff, assumed to be professionally more competent and efficient, and who can take the burden off the shoulders of field staff, is needed. It is also necessary to have some degree of independence from project management to allow it to perform critically and independently. This especially applies in determining the relevance of project objectives and, hence, the validity of some of its activities. The unit also requires freedom to achieve its objectives. However, too much of independence can have a negative effect and lead to alienating the project management staff.

Indeed, the introduction of units having the responsibility for the design and implementation of M&E systems has always underlined the need to involve field personnel more closely at all stages. Since the purpose of M&E is to assist field implementers to function more efficiently and effectively, this aspect is being increasingly recognized as very desirable.

AGENCY FOR M&E

Regarding the question on who should carry out the monitoring, Bahuguna and Upadhyay (2004) suggested a framework and a few parameters that can be adopted by the NAP with suitable modifications (Box 1.2).

M&E AND INFORMATION PROCESSING

The M&E exercise produces information of varying accuracy and validity. The direct measurement of physical variables, such as nurseries established, number of seedlings raised, and number of seedlings planted, is likely to be most accurate. Indirect measurement of sample variables using socio-economic surveys will certainly involve sampling and measurement errors (arising from many sources, including faulty recall of respondents). Informal contacts increase the chance of biased sampling (group leaders may not be representative). Moreover, it is not easy to summarize from unstructured interviews a sample of judgements and opinions, or be sure of its validity. M&E units can give users some idea of the accuracy of their results. If they give the impression that they are of equal accuracy, some very wrong decisions could result.

Box 1.2 | Monitoring by whom and why?

Monitoring needs to be done at various levels. State forest departments should carry out internal monitoring at all levels. At the national level, it can be done by the National Afforestation and Eco-development Board (NAEB) and another board at the state level. Monitoring would be done mainly with the help of feedback loop and monitoring reports received from field agencies by the state nodal officers.

The task of external monitoring can be entrusted to any independent agency by the state or the central government. The agency could be a non-governmental organization (NGO) or an academic or research institution with proven and unbiased track record.

At the local level, village forest protection committees can be made responsible for monitoring the programme. NGOs and self-initiated groups can also assist the committees or independently conduct studies on the evaluation of the programme. Participatory monitoring can be encouraged to be an integral component of monitoring in the states to ensure the elements of ownership.

Evaluation

Evaluation studies can also be undertaken by the state and central governments through independent agencies and also by NGOs, academics, and research organizations. They need to adopt appropriate and standardized methodologies for data collection and analysis ensuring reliability and scientific validity of the monitoring. Premier forestry and statistical research organizations can be entrusted with the task of developing standardized data formats, statistical sampling, analysis and interpretation techniques, and a framework for reporting. The NAEB constantly refines this with the assistance of many national-level institutions such as the Indian Institute of Forest Management. So far, not many studies have been undertaken in a systematic manner and no benchmarks for impact assessment exist. Some initial studies can serve as baseline data for subsequent monitoring.

Source Adapted from Bahuguna and Upadhyay (2004)

RELUCTANCE IN INCORPORATING M&E

It is fashionable now to incorporate project M&E without any question. Practically, many M&E systems are found to be poorly operated and of limited effectiveness, even though they may have been well designed and run by competent staff. M&E activities face unstated opposition or covert resistance for several reasons. Sometimes, there is an antipathy towards monitoring because it exposes deficiencies in project implementation or, on some occasions, it reveals weaknesses in project management. At times these can be linked to individual members of project staff, and when they are punished, obviously they grudge the system. In some cases, the staff devote a great deal of their time in monitoring the activities, which diverts their efforts from the day-to-day running of a project; this is sometimes resented. Again, much of the monitored output may not be used by the implementing staff because they may not have been consulted about the monitoring system and, therefore, may have little idea of what the information is for. Understandably, they would assume that it would be of little value to them in the performance of their project duties.

Hence, a conscious effort must be made to deal with the reasons for this resistance. An effective way of doing this is to have a close liaison between implementing staff and the M&E unit, so that the former will understand the purpose of the system, not only to provide for management needs and improve its efficiency, but also to improve the relevance and impact of the project.

OPTIONS FOR THE FUTURE

Various dimensions of M&E have been discussed, which are necessary for any forestry/afforestation project. Some of the parameters of technical, social, economic, and political nature are very critical from the forestry perspective, and accordingly they have been highlighted. The NAP has got most of these elements in the comprehensive M&E framework, as is evident from the current operational guidelines of the NAP for the Tenth and Eleventh Five-year Plan periods. There are still some limitations in implementing the M&E systems. Some findings of M&E indicate that corrective actions have to be taken. In this light, the present chapter and particularly Chapters 5 and 6 are very useful to all stakeholders involved in the aspects related to the M&E of the NAP.

ACKNOWLEDGEMENTS

This chapter is derived extensively from many reports (mentioned in references) and mainly from the FAO Forestry Paper 60, titled "'Monitoring and evaluation of participatory forestry projects 1985" and the author sincerely acknowledges the Food and Agriculture Organization and the contributing authors of the aforementioned reports.

REFERENCES

Bahuguna, V. K. and A. Upadhyay (eds). 2004. *Root to Canopy: monitoring needs for JFM—the perspective of policy makers*. India: Winrock

Bhattarai, T. N. and J. G. Campbell. 1984. Data collection guidelines for monitoring and evaluation of community forestry activities in Nepal. HMG/UNDP/FAO Community Forestry Development Project, Nepal.

Casley, D. J. and D. A. Lury. 1981. A *Handbook on Monitoring and Evaluation of Agriculture and Rural Development Projects*. Washington, DC: IBRAD

Clayton, E. and F. Petry. 1984. *Monitoring Systems for Agricultural and Rural Development Projects*, Vol. 1 and 2. Rome: FAO

FAO (Food and Agriculture Organization). 1985. Monitoring and evaluation of participatory forestry projects. *Forestry Paper No. 60*. Rome: Policy and Planning Service, FAO

IFAD (International Fund for Agricultural Development). 1984. *Guiding Principles for the Design and Use of Monitoring and Evaluation in Rural Development Projects and Programmes*. Rome: ACC Task Force, Monitoring and Evaluation Publications

Slade, R. H. and R. Noronha. 1984. *An Operational Guide to the Monitoring and Evaluation of Social Forestry in India (Working Draft)*. Washington, DC: World Bank

CHAPTER 2

Strengthening Evaluation and Monitoring Mechanism under National Afforestation Programme

S K Barik

INTRODUCTION

Monitoring and evaluation (M&E) hold the key to the success of any programme. The evaluation of a programme generally demands two forms of measurement: (1) measurement on an internal scale and (2) comparison against some external standard (Treweek 1999). At present, the evaluation of the National Afforestation Programme (NAP) is largely restricted to measurement on an internal scale only (NAEB 2002). There is, therefore, a need to evaluate the NAP (at least at the end of the plan period) based on the second form of measurement, that is, comparing the performances with other similar programmes, preferably with a similar externally aided project.

There has been a lot of debate over efficiency and validity of different techniques used for evaluation. Difficulties arise when qualitative and quantitative factors are combined. One of the options is the use of multiple evaluation criteria, which are then given weight or priority using a scoring system (Usher 1986). In the instances where objective scoring systems are to be combined with more subjective weighting, it is important that there is an absolute clarity about (1) what component of the programme is being evaluated, (2) what criteria are to be used, and (3) how they are to be measured. No standard set of criteria has emerged, and multitudes have been used in practice. Most criteria are also difficult to apply consistently, because all of them cannot be measured using defensible, consistent, and objective techniques. For example, although it is possible to derive objective and numerical estimates of "survival and growth rates of planted seedlings", it is much less straightforward to measure attributes such as "people's participation" under the programme. At a more strategic level, it is useful to determine the impact of NAP on achieving the national target of 33% forest and tree cover by the end of the next plan period. In addition to estimating the extent of contribution that the NAP is making towards achieving the national target of natural resource sustainability, there is a need to identify the properties that affect the continuity/long-term viability of the programme itself. The evaluation framework must include both self-evaluation and external independent evaluation. Several authors have argued that self-evaluation and independent evaluation are mutually complementary components of the evaluation system.

Monitoring entails the systematic observation and measurement of the programme components to establish their characteristics and changes over a period of time. In contrast to evaluation, which is a one-time assessment of the project performance in relation to the stated project objectives (mostly undertaken at the end of the project), monitoring is a continuous process along a time gradient. The key parameters for programme monitoring need to be carefully developed to make monitoring effective.

This chapter critically reviews the M&E process being currently followed in the NAP (NAEB 2002) and suggests measures for improvement in the overall framework to make the M&E more effective and beneficial.

REVIEW OF THE CURRENT PRACTICE OF M&E UNDER NAP

The evaluation and monitoring framework of NAP at present consists of three levels, each level being complementary to the other.

Level 1

The chief executive officer and chairperson of the concerned forest development agency (FDA) should regularly monitor the day-to-day activities of the plantation site on each day of their visit. It is mandatory for them to write in the plantation journal their comments, observations, and suggestions for improvement of work. The next level of responsibility for internal monitoring rests either with a monitoring committee consisting of officers in the rank of chief conservator of forests (CCF) and above, constituted at the state level, or with senior supervisory officers such as the CCF, additional principal chief conservator of forests, and principal chief conservator of forests. The monitoring committee submits its report from time to time. The inspecting officers submit their inspection notes, which can also be treated at par with the formal monitoring report.

Level 2

The second level of M&E is undertaken during the project period by an independent organization, appointed by the chairperson of the concerned FDA. The funds for this purpose are provided by the National Afforestation and Eco-development Board (NAEB) as an integral part of the project under the head "monitoring and evaluation". The report is submitted by the independent evaluator to the FDA for onward transmission to NAEB.

Level 3

The third level of M&E is conducted by the NAEB through its appointed agency. Such evaluations are undertaken twice in a project cycle—one as a mid-term evaluation at the end of the second year of project implementation, and the other one on the completion of the project. Besides these institutionalized M&E mechanism, officers of the NAEB and its regional centres also make random visits to the NAP project sites.

The present M&E mechanism adopted by the NAEB seems to be adequate, and the manual in use addresses most issues fairly in detail. The manual has ample scope to be descriptive on specialized issues. The procedure adopted is, by and large, in conformity with international frameworks of M&E. However, to bring more objectivity to the M&E process, the following modifications in the existing manual may be considered.

- The format needs to be more precise and objective and should be based on more quantitative indicators.
- The indicators should not be too many.
- More sensitive indicators should be explored.
- The indicators must be practical and cost effective.
- The indicators need to be both quantitative and qualitative.
- The importance of experience, good guess, and proxy indicators should not be undermined.

The most important aspect in the M&E process, which has not received adequate attention in the present-day framework, is the utilization of M&E results. The M&E process should not end at data collection. The data should be critically analysed for focused learning and introducing future corrections. The subsequent paragraphs describe steps followed by several international agencies in "results-based monitoring system", which can help in achieving this objective.

STRENGTHENING PROJECT PROPOSAL

Both "quality at entry" and "quality at exit" are essential aspects for the success of any project. The NAP seems to be logically concerned about the latter. However, the quality aspect at the launch of the project has been largely neglected. Adoption of a logical framework (LOG FRAME) during the project formulation should be encouraged as prescribed in the NAP guidelines (Box 2.1). During the application of LOG FRAME, the overall objective and impact of NAP at the country level may remain the same, but the purpose, outcome/output, and activities should be site, state, and project specific.

The objectively verifiable indicators (OVIs) and means of verification (MOVs) (Box 2.2) should be unambiguously identified during the project formulation stage. These, in fact, provide the basis for the project M&E system. It may be noted that the OVIs and MOVs should be practical and cost effective, and the indicators should not be too many.

BOX 2.1 | Objectives in the LOG FRAME for project formulation

Hierarchy of Objectives in the LOG FRAME

- **Super goal**
- **Goal:** higher order of objective to which the project contributes.
- **Purpose:** the project-level effect or impact of project.
- **Results:** the deliverables of the project or the terms of reference.
- **Major activities:** the main activities to be undertaken to accomplish results.

Explanation of the Terms Used in the LOG FRAME

- **Goal:** Why is the project carried out?
- **Purpose:** What the project intends to achieve?
- **Results:** What results the project must bring about?
- **Major activities:** How is the project going to achieve these results?
- **Assumptions:** What external factors are crucial for project success?
- **Indicator:** How should the success of the project be assessed?

Box 2.2 | OVIs and MOVs for project M&E

Objectively Verifiable Indicators (OVIs)

- Goal
- Purpose
- Results
- Major activities

Means of Verification (MOVs)

- Source of data
- Needed to verify
- Status of results
- Level of indicator

ADOPTION OF RESULT-BASED MONITORING SYSTEM

Under a result-based monitoring system, the success of projects should be measured not only by the services delivered but also by the objectives achieved (GTZ 2004). In such a system, all monitoring activities of a project or programme are geared to observe results. It forms part of the self-evaluation as well as the independent evaluation of a project/programme. The result may be intended or unintended, expected or unexpected, and positive or negative. Various tasks must be performed to identify these results. The term "result" is only used to denote those changes that can be attributed to a project/programme. Mere occurrence of a change is not sufficient to attribute it to a project or programme. Only where a causal or, at least, a plausible link can be established between the observed change and the programme activities, such results can be considered project/programme results. While designing a result-based monitoring system, one would find that many monitoring parameters do not differ fundamentally from one project to the other. For example, the following parameters remain unchanged irrespective of the project location and size.

- What the objective is and how a project or programme would achieve its objectives?
- How a project/programme can identify the development results that it has helped to generate?

Results-based monitoring analyses the monitoring data using the whole results chain from inputs, via activities and outputs to outcomes and impacts

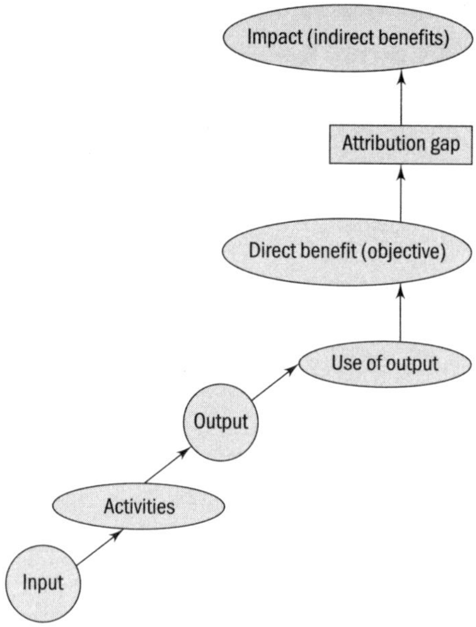

Figure 2.1 *The result-chain model*
Source GTZ (2004)

(Figure 2.1). The distinctive feature of this system is not only its focus on what has been done, but also its attempts to identify changes generated by the project activities. A detailed description of the result-chain analysis is as follows.

The activities that generate the outputs are launched using various project components as inputs. These are then utilized by the target groups or intermediaries (use of outputs), generating medium-term and long-term development results, that is, outcomes and impacts. Attribution is relatively easy in most cases up to the level of "use of outputs". As we climb up to the levels of "outcomes" and "impacts", the external factors that cannot be influenced by projects and programmes become increasingly important. The attribution gap widens up to an extent where the observed changes are not directly related to the project outputs any more. Development changes that can be shown directly are termed "direct benefits", and the project is entitled to claim such observed positive development changes, that is, the direct benefits. The project and programme objectives are set at this level. As mentioned above, projects and programmes aim to generate impacts beyond the objectives, including a host of "indirect benefits". These are usually the ultimate reason for the intervention. However, it is not possible to identify a causal relationship explaining how these indirect benefits came about, as many actors are involved and the interaction of many factors brings about such benefits. Thus, it is practically impossible to isolate the effect of a single intervention.

However, we must identify the outcomes that are clearly attributable to the project/programme and are defined as the project/programme objectives. On the other hand, it is also necessary to identify those impacts that can no longer be clearly attributed to the programme/project, although generated in the wider project/programme environment in the wake of the achievement of objectives.

Thus, the task of project/programme monitoring includes analysis of the M&E data along the aforementioned lines and establishing what contribution the project/programme might have made towards the observed changes. While undertaking such analysis, it is not necessary to examine the basis of the monitored data on inputs, activities, outputs, use of outputs, and outcomes. Rather, the analysis should aim at answering how the project/programme might have contributed towards these changes in its environment.

SUGGESTED IMPROVEMENTS IN THE PRESENT M&E MANUAL

Methodological Modifications

For adopting result-based monitoring, the existing instruments need to be further refined and supplemented. Some suggested modifications are as follows.

Decentralized and participatory evaluation

Special effort should be made to revise and develop new evaluation procedures in compliance with the principles of decentralization. Thus, the evaluation should be performed by a team consisting of officials from the forest department, FDA, joint forest management committee (JFMC), and the evaluator. The list of names of each participant category should invariably be mentioned in the report, because the set of individuals/groups involved in the M&E is a key indicator of the methodological direction.

Data collection and survey method

While deciding the data collection and survey method to be used, its costs and benefits should be considered. Group discussions, participatory monitoring, action research, participatory rural assessment, and similar qualitative methods provide information from a subjective perspective, which is often more revealing than purely statistical data.

Keeping to the essentials

Essential variables and indicators should be carefully chosen to avoid the collecting data on many less important and less cost-effective variables/indicators.

Interpretation of M&E data

Although graphs and tables are helpful, data interpretation is the key step towards utilization of M&E data. Adequate attention should be given towards this aspect.

Limiting reports to the essential

The communicative potential of lengthy reports is limited. Hence, reporting should be limited to the essentials.

Monitoring as positive feedback to NAP

The monitoring system should lead to "lessons learned" and be a positive feedback to NAP.

Sampling size

- At least 20% of the area to be covered under each component.
- At least 10% of the total JFMC to be covered.
- At each plantation site, random quadrats or belt transects of 0.1 ha (10 × 100 m) should cover at least 10% area to be taken as sample at each site.
- Species-wise collar and height measurement of at least 50 individuals from each species.

Analysis of evaluation reports

- The analysis should be based on the result-chain model.
- It could be a part of the M&E format or may be taken up separately during analysis of the report by the NAEB.

Table 2.1 gives some examples to modify the format of evaluating reports.

TABLE 2.1 Examples to modify the format

Project attribute	*Indicator*
Overall project attributes	• Physical • Financial • Sustainability (of institutions, assets, and programme)
Physical works	• Survival • Growth of saplings • Maintenance (weed density, social/biological fencing) • Soil and moisture conservation • Entry-point activities
Functioning of JFMCs	• Degree of people's participation • Involvement of women • Decision-making power of weaker sections • Sense of ownership • Satisfaction over perceived benefits
Functioning of FDA	• Composition • Effectiveness in administration, monitoring, and supervision • Flow of funds • Planning, implementation, and maintenance of assets by JFMCs • Meetings, trainings, and capacity development for alternative income generation
Poverty alleviation and ecosystem services	• Employment generation • Alternative income generation • Poverty alleviation through project activities • Ecosystem services • Empowerment

FDA – forest development agency; JFMC – joint forest management committee

INDICATORS

The indicators for the evaluation of NAP may be one of the following types: quality, quantity, time, place, and beneficiaries. It is advisable to use a combination of quantitative and qualitative indicators depending on the type of variable being studied. For instance, qualitative indicators may be converted to quantitative indicators by using a 5 or 10 point graded scale. To bring uniformity, all the quantitative indicators may also be converted to the same scale. As a result of such simple exercises, objectivity in the assessment of a number of variables can be enhanced. The overall grading would then be more objective. It may be noted here that although efforts should be made to quantify as many variables as possible, quantification in no way undermines the importance of subjective assessment for many variables.

AWARDING FINAL GRADE

Using the appropriate indicators, the final grade may be worked out, which is shown in Table 2.2.

TABLE 2.2 Indicative final grading table (example of M&E data)

Criterion	*Indicator*	*Quantitative value*	*Quantitative grading*	*Qualitative grading*	*Overall grading*
Overall	Survival (%)	67	6.7	—	6.7
	Growth (cm)	34	5.6	—	—
	Sustainability	—	—	7.5	7.5
Physical works	—	—	—	—	—
JFMC functioning	—	—	—	—	—
FDA functioning	Impact of training	—	—	5.0	5.0
Poverty alleviation and ecosystem services	—	—	—	—	—
Overall	—	—	—	—	—

FDA – forest development agency; JFMC – joint forest management committee

CONCLUSION

The suggested changes should enable the M&E framework to derive meaningful information, which can be put to use for project/programme management. The ultimate objective of the modified M&E system should be to deliver the "lessons to be learned" as clearly as possible.

ACKNOWLEDGEMENTS

This chapter derives information from several technical reports, particularly those prepared by GTZ. The author thankfully acknowledges these sources.

REFERENCES

GTZ (Deutsche Gesellschaft für Technische Zusammenarbeit). 2004. *Results-based Monitoring: guidelines for technical cooperation projects and programmes.* Eschborn: GTZ

NAEB (National Afforestation and Eco-development Board). 2002. *National Afforestation Programme Guidelines*. New Delhi: NAEB, Ministry of Environment and Forests, Government of India

Treweek, J. 1999. *Ecological Impact Assessment*. Oxford: Blackwell Science

Usher, M. B. 1986. *Wildlife Conservation Evaluation*. London: Chapman and Hall

CHAPTER 3

Critical Assessment of Methodology for Monitoring and Evaluation of NAP Projects

J Kulkarni

INTRODUCTION

Almost from the beginning, the National Afforestation and Eco-development Board (NAEB) has been carrying out rigorous evaluations of its afforestation projects with a view to improving performance. This shows the importance it gives to evaluations by external agencies. Its evaluators have been selected with great care, and most of them have substantial experience in the field, which has resulted in generally high quality of evaluations. The importance given to the quality of evaluations puts substantial responsibility on the evaluators; they need to justify the confidence placed on them by the NAEB to give objective, incisive evaluations.

OBJECTIVES OF EVALUATION

Evaluation of projects fulfils several objectives. First, it helps in assessing the degree of achievement of project objectives and areas in which there are shortcomings. If the evaluation is carried out properly, it can help identify the reasons for non-achievement and systemic faults that prevent achievements of objectives. Second, a good evaluation by pointing out the faults and making practical recommendations can help improve project performance. These improvements are possible, provided the evaluation findings are communicated to the implementing agency.

A major function of evaluation is to give feedback to the funding agency, in this case the NAEB. Taking cognizance of negative evaluations, the NAEB can bring the matter to the state forest department, pull up the concerned forest development agency (FDA), and ask it to improve its performance. If the desired action does not take place, the NAEB can, under extreme circumstances, decide to end the project with that FDA. In the case of FDAs with good evaluations, the NAEB can take positive action by citing them as examples of good performance, which can be emulated by other FDAs.

Through a series of workshops, the NAEB is preparing the ground for further improvement in the quality of evaluations by carrying out a critical evaluation of its methodology. The objectives of this are as follows.

- Improve the quality and reliability of inputs from various stakeholders.
- Improve the objectivity of reporting analysis.
- Improve the follow-up action by implementers and supervisors.

It is obvious that the aforementioned objectives have originated out of the need felt by the NAEB to rectify drawbacks commonly found in the evaluation reports themselves and the process of follow-up and action on the evaluations by respective state governments. The evaluation process has been broken into a number of components here with each component discussed and improvements suggested in the light of these objectives.

EVALUATION FORMAT

The NAEB has been carrying out evaluation in a particular format, which was created when evaluation of projects started for the first time. This format has undergone changes leading to some improvements, but its basic structure has remained the same. The format covers almost all the important points required for evaluation. However, the grouping used in the evaluation structure is slightly confusing and does not lend itself to easy writing of evaluations. Closely related topics have been placed in different groups making their collective discussion difficult.

These points need to be regrouped under different heads for proper sequencing of evaluation matter and organization of the report. Box 3.1 summarizes the existing format.

Box 3.2 makes suggestion for a new evaluation structure essentially covering the same points.

GRADING SYSTEM

Projects are, at present, rated according to a grading system ranging from 1 to 10 in four categories: outstanding,, very good, good, and poor. Thus even if a project receives 3 to 5 points out of 10, it is still graded as good, which

BOX 3.1 | Existing structure of evaluation format

General: Basic information about the project area.

Quantitative: Information about target/achievement, people's participation, entry-point activities, joint forest management, capacity building, monitoring and evaluation, and forest development agency mechanism.

Qualitative: Site selection, survival rate, plantation parameters, people's participation, protection levels, maintenance of records, project benefits, project reporting activities, constraints/limitations, suggestions for improvement.

BOX 3.2 | Suggested structure for evaluation format

Project area: Basic information about the project area, location, topography, soil, rainfall, vegetation, people, biotic pressures, need for afforestation.

Project information and design: Project proposal, sanctioned proposal, details of project proposals and sanctioned proposal.

FDA mechanism: Forest development agency (FDA) structure and membership, FDA meetings, communication between FDA and the joint forest management committee (JFMC).

Project management and effectiveness: Achievement of physical and financial targets, information availability, record keeping, project reporting, inspections by senior officers, project evaluations, fund flow, fiscal management adopted for motivation of people.

People's participation: Joint forest management issues, entry-point activities, JFMC meetings, women's participation, participation of weaker sections and equitability, people's involvement in decision-making, choice of species by people, people's satisfaction with project, perceived benefits of project, people's willingness to protect assets, benefit-sharing mechanism, people's awareness, and approach adopted to motivate people.

Forestry and technical issues: Site selection, choice of species, technology used, plantation techniques, fencing methods, suitability of model, survival rate, health of plants, and nursery techniques.

Capacity building/awareness/motivation: Type of trainings and exposures, visits, subjects, methodology used, trainers, number of trainings, number of staff and JFMC members trained, fund utilization.

does not appear to be reasonable. A revised, more rational grading system has been suggested. Both the existing grading system and the suggested one are given in Table 3.1. Having said this, it must be stressed that the grades received by a project are not the most important part of the evaluation. The analysis of the project, identification of shortcomings, and recommendations for improvement are more important than the grade received by a project.

The points received by a project may be decided using the same structure employed in evaluation of the project. Table 3.2 shows the procedure for calculating points received by a project if the evaluation structure suggested in this chapter is followed.

Suggested weights for each section are given in the second column, but the evaluator can decide his or her own weights. Marks received (weighted average) = Y/X.

The grading by the evaluator should be as objective as possible. Subjectivity cannot be eliminated totally. However, it can be reduced by providing justification for marks given under each item.

STEPS IN EVALUATION

An evaluation is likely to progress smoothly if the evaluator follows a systematic approach. The main stages are described as follows.

TABLE 3.1 Suggested changes to grading system

Present system		*Suggested system*	
Points	*Grade*	*Points*	*Grade*
8 to 10	Outstanding	9 to 10	Outstanding
5 to 8	Very good	8 to 9	Very good
3 to 5	Good	6 to 8	Good
<3	Poor	4 to 6	Average/satisfactory
		2 to 4	Poor
		<2	Very poor

TABLE 3.2 Weighted point system for determination of overall points

Item	*Weight*	*Points (out of 10)*	*Weighted points*
A	*B*	*C*	*D = B × C*
Project design	1		
FDA mechanism	2		
Project management and effectiveness	2		
People's participation	4		
Forestry and technical issues	4		
Capacity building/awareness	2		
Total	X		Y

The first step involves contacting the concerned FDA and informing it about the details of the appointment for carrying out the evaluation of that FDA. Dates for the evaluation visit can be finalized after conducting a discussion with the member secretary/chief executive officer (CEO) of the FDA. It is very useful to send formats for data collection in advance so that the FDA can keep the required information ready before the evaluation visit. Generally, an experienced evaluator should make his or her own formats for information collection and ask the FDA to fill in the information in these formats. Proper formats are convenient for the FDA to provide the required information. Lack of formats or improper formats will cause delays and mistakes in the information provided by the FDA.

In the initial meeting, the evaluator should convey to the FDA member secretary/CEO his or her information requirements and his or her evaluation strategy. After this, the evaluator should obtain the basic information for the evaluation. It is important to have the required information before beginning the evaluation. An efficient FDA will keep the required information ready or provide the evaluator with a project summary giving salient features of the FDA. The evaluator should have a discussion with the FDA member secretary/CEO about the achievements of the project, the approach adopted by the FDA, forestry and joint forest management (JFM) aspects of the project, and the problems faced in the implementation of the project. After satisfying that all the information necessary for carrying out the field visit is available, the evaluator should finalize the tour itinerary in consultation with the member secretary/CEO.

The field inspection will normally consist of inspection of plantations and other assets created during the project life. It will also involve interactions with the joint forest management committees (JFMCs) associated with the project. This is the most important part of the evaluation and is discussed in detail in the subsequent sections.

The evaluator should have a wrap-up meeting with the FDA member secretary/CEO after the field visit. During this meeting, he or she should obtain clarifications from the member secretary/CEO about various observations made during the field visit. If there are any negative findings about the project, the evaluator should point them out and discuss them with the member secretary/CEO and obtain justification, if any. The evaluator should also collect all required information about the project. Some of the information may not be available immediately, and the FDA may have to collect them from the field. If the information cannot be obtained immediately, the evaluator should make sure that the exact information required is understood and will be sent soon by the FDA.

ESTIMATION OF SURVIVAL PERCENTAGE

An important section of the forestry part of the evaluation is the estimation of survival percentage. The NAEB has not yet issued definite guidelines on estimating the survival percentage for National Afforestation Programme (NAP) evaluations. It, however, has issued guidelines for this purpose for another activity, that is, nationwide estimation of survival percentage of all afforestation activities under its Twenty Point Programme through sample counts. These guidelines may be used as a starting point for estimation of survival percentage during evaluations.

The guideline states that 10% of the total area of plantations should be sampled (Box 3.3). Generally, this target can be achieved easily. The area of plantations will differ depending on the stage at which the evaluation is carried out. In the case of first evaluation, only 40%–60% of the target may have been completed. The evaluator should try to cover at least 10% of the total approved project area.

The selection of plantations should be carried out in a stratified random method. This implies that the evaluator should stratify the plantations. Criteria for stratification can be (1) model, (2) forest range, and (3) year of plantation. So there are already three levels of stratification. Generally, more than one level of stratification is not advisable. It would be proper to carry out stratification on the basis of the model since this is the most striking criterion. However, stratification has meaning only if survival percentage is estimated by a statistically sound procedure. Otherwise, it does not have much significance.

BOX 3.3 | Guidelines for estimation of survival rate through sample counts for afforestation work under the Twenty Point Programme

Ten per cent of the total area under plantations has to be sampled for survival rate studies with each sampling unit of size 0.1 ha. This is to be stratified on the basis of areas planted under different schemes by the implementing agencies. The samples taken should be unbiased and random. The statistics should conform to +5% of standard error of mean.

In practice, strictly random sampling of plantations may not be possible because of the practical constraints of approachability and planning of a convenient route to sample maximum plantations. Random sampling is, therefore, an ideal, which the evaluator should try to achieve to the extent possible.

The current guidelines state that the sampling unit should be 0.1 ha, but they do not state the sampling intensity, that is, the number of plots to be sampled in a plantation. They state that statistics should conform to +5% of the standard error of mean. Since most evaluators are not statistical experts, it is very difficult for them to carry out these statistical calculations. The NAEB should consult statistical experts and evolve a statistically sound methodology for evaluation of survival percentage. These guidelines should be drafted in the form of a monograph and circulated to all the evaluators; it should include statistical equations, statistical tables, and solved examples. Till then, evaluation of typically 5%–10% sampling is likely to be sufficient.

INTERACTIONS WITH JFMCS

People's involvement is of paramount importance in protecting forests and for successful implementation of projects. Even if project implementation is technically perfect from the forestry point of view, the project will not achieve the desired objectives if the JFMCs do not participate wholeheartedly in the project. Interacting with JFMCs and obtaining their feedback are perhaps the most important aspects of the evaluation.

At the beginning of the evaluation, it is necessary to inform the FDA authorities that the evaluator would like to meet the JFMCs. The interactions with JFMCs should be held, as far as possible, at a convenient time for the members, such as early morning or evening. If the programme is well planned from the beginning, the exact date and time can be fixed in advance so that the JFMC members have advance intimation of the evaluators' visit. The JFMC office-bearers should be present. Other members should also be present in sufficient numbers so that all the viewpoints can be heard. The evaluator should also insist that women members be present in the meeting. The JFMC member secretary should be prepared with JFMC records such as the microplan, resolution/proceedings register, bank passbook, cashbook, and plantation journals.

The evaluator should be prepared with a checklist or datasheet of information to be collected to ensure that information on each important item is collected. There should be flexibility in the interactions so that any point found to be interesting or important can be pursued in detail while unimportant points can be disposed of quickly.

RELIABILITY OF INPUTS FROM STAKEHOLDERS

Reliability of inputs from various stakeholders is of utmost importance because it is the foundation of the evaluation. Unreliable or wrong inputs lead to information bias and can compromise the quality of the evaluation. Lack of reliability can be of two types: (1) wrong or misleading information and (2) withholding information. Although not a deliberate act all the time, withholding information can surely reduce the quality of an evaluation, as the evaluator may remain unaware of some useful information on a particular aspect of the project.

The NAEB has already made it a practice that important information about FDA activities, such as reporting to the NAEB, FDA meetings, receipt of letters, and writing of letters regarding requisition for funds, should be verified with FDA records. This ensures, to a great extent, the reliability of the information received from the FDA.

Information bias can be minor, such as trying to portray a rosier picture of project performance than actual facts. Serious information bias can occur as a result of misleading and false information given by an interviewee. Obtaining absolutely objective information from JFMCs and field staff cannot be guaranteed. There is a human tendency to portray facts in a good light and hide shortcomings. For example, the JFMCs will always say that they are cooperating fully with the FDA staff to protect plantations. The field staff, on the other hand, will try to blame the JFMCs for failures. A good evaluator should be able to sift through the inputs and determine the reality.

Some causes of bias in information received from JFMCs and field staff are as follows.

- Show facts in a good light (all).
- Hide shortcomings; avoid responsibility/accountability (all).
- Fear of official machinery (JFMC members).
- Fear of reprisals (JFMC members/FDA staff).
- Inter-group or personal rivalry (JFMC members/FDA staff).

An evaluator interested in carrying out an in-depth and factual evaluation needs to be alert to these possibilities and should try to reduce them to the extent possible. Some ways of reducing bias are as follows.

Cross-checking Information

First, information bias can be reduced by cross-checking with multiple sources. For example, if doubtful information is received from a JFMC member, it should be cross-checked with other JFMC members and the concerned forest department staff. If this is not satisfactory, the information may be cross-checked with the FDA member secretary/CEO. Second, information bias can be reduced by cross-checking with official and secondary data. It requires experience on the part of the evaluator to be able to spot discrepancies in information given by various sources.

Confidential Interviews

In spite of all the efforts, it is not always possible to eliminate certain types of biasness. It is particularly difficult to avoid information being withheld, which is generally due to the fear of reprisals on the part of JFMC members and sometimes on the part of FDA staff. This is understandable because JFMC members are dependent on field staff for many small favours such as permission to collect firewood, bamboo, and non-timber forest products from the forest. They would normally not like to antagonize the FDA staff in any way by making negative reports against them.

Circumstances that are likely to discourage free and frank communications on the part of JFMC members are public gatherings and the presence of FDA staff. A confidential setting and one-to-one interviews are likely to result in more information sharing on the part of the interviewees. Therefore, confidential interviews are suggested as an adjunct to the existing methodology to increase reliability of information collected from JFMC members.

It is not normally possible to hold confidential interviews during the first round of inspections, since the evaluator will normally be accompanied by FDA officers and staff, and the time available for such an exercise will be limited. It is recommended that the evaluator shortlists a few JFMCs after the first round of inspections and carries out independent surveys conducting confidential interviews with JFMC members in these villages. While carrying out these surveys, the evaluator should follow some kind of stratification system. The following system is suggested.

- JFMCs selected for survey (1 per range).
- JFMC executive committee members (1–2 from each JFMC).
- House-to-house survey representative of various strata—castes, gender, and income group (10–20 per JFMC).
- Forest department staff (range forest officer, forester, and guard).

There may be some practical difficulties in carrying out surveys involving confidential interviews. However, such a practice is likely to improve the reliability of the inputs from JFMC members.

It is advisable to carry out the confidential interview survey using structured questionnaires. Some suggestions for the design of the questionnaire formats are as follows.

- The questionnaire should be brief and focused. If the questionnaires are long, the interviewee will lose interest.
- The questionnaire should test the interviewee's knowledge of the project. Since some of the interviewees may be illiterate, they may not know the quantitative details of the project. In such cases, knowledge of the objectives of the project and its basic features should be considered satisfactory.
- The interview should elicit information on both positive and negative aspects of the project.
- The questionnaire should have sections for technical and social issues.
- It should cover gender-related issues, especially the needs of women if they have been addressed in the project.
- It should cover likely shortcomings in implementation and ask for suggestions in improving the project.

The inputs obtained from confidential interviewees may also be biased at times because the interviewee may deliberately provide false information. Such cases are likely to be rare. If there is any doubt about the information obtained, the evaluator should try to cross-check with other interviewees and information sources to the extent possible. In such situations, the evaluator should exercise caution and use his or her judgement while making conclusions. The information obtained from confidential interviews is not a substitute for other aspects of evaluation.

OTHER EVALUATION PARAMETERS

Evaluation of the project should be carried out using several parameters. The evaluator needs to be alert for information that may reveal either the strong points or the shortcomings of the project. Therefore, while evaluation can be carried out in a systematic and planned manner, it cannot be carried out in a standardized manner. Experience and alertness of the evaluator are very important. Some aspects of project management quality that can act as indicators for an evaluator to judge the project are as follows.

- The availability and quality of information provided to an evaluator are indicators of efficient management of a project. Well-managed FDAs often have information readily available in their records. Often, most of the information required is provided to an evaluator in the form a concise report. On the other hand, it is difficult to get necessary information easily in poorly managed FDAs.
- Measures of innovation adopted by FDAs are also an indicator of project performance. These can be in any component of the project such as in plantation techniques, soil and moisture conservation works, entry-point activities, people's involvement, record keeping, accounting systems, training, awareness programmes, and exposure visits.

OBJECTIVITY OF EVALUATION

Subjectivity in evaluation arises when the evaluator relies on feelings and impressions to form an opinion about the project. Every evaluator forms impressions, but an evaluation is objective when the impressions are based on verifiable facts. Reliability of information is very important, which has been dealt with in previous sections. To ensure objectivity, the evaluator should justify the opinions formed and comments mentioned about the project with definite factual statements.

The NAEB has already made a valuable beginning in ensuring objectivity of observations by conducting regional workshops where evaluations are presented before FDAs and forest officers. Evaluators have to weigh their evaluations carefully and justify their observations in the presence of FDA officers and cannot get away with loose statements. A great deal of objectivity is, therefore, ensured by these interactive workshops.

To justify the comments made, the evaluator should be capable of citing specific observations about the project. The observations may be either positive or negative. Categories of such observations are as follows.

- Technical aspects of plantations and soil and moisture conservation (SMC) works, appropriateness of technology used, strong points or mistakes in implementation of plantations and SMC works, and protection levels.
- Specific observations about functioning of JFMCs, relationship of JFMC members with FDA staff, attitude of JFMCs towards plantations and forests, JFMC record keeping, and knowledge and awareness of JFMCs about JFM and the project.
- Financial management of the project, FDA functioning, and FDA meetings.

If an evaluator makes a number of specific observations about a project, this gives a good idea of the quality of the project. This is far more effective than merely saying that the implementation of a project is good or bad. Further, an evaluator should be able to identify deviations from the norm in various aspects of the project—JFMC functioning, field plantations, or FDA functioning. For this, an evaluator should be aware of the NAEB guidelines and norms for NAP plantations. If there is a deviation from norms or guidelines, it should be discussed with the member secretary/CEO and their justifications should be recorded.

NAEB ACTION FOR IMPROVING QUALITY OF EVALUATIONS

The NAEB has already made considerable efforts to improve the quality of evaluation, first by holding regional workshops for presentation of evaluations, and second by critically evaluating the methodology in the workshops. Considerable inputs and feedback have already been made available to the NAEB through these means.

Each evaluator or agency carries out the respective evaluations differently. A workshop consisting mainly of evaluators would allow the evaluators to benefit from each other. Such an opportunity should be provided by the NAEB. Case studies can be presented, and various procedures and methodologies can be discussed. Brainstorming sessions during such workshops may engender some innovative ideas for improving standard of evaluations.

Taking the matter further, an attempt should be made to standardize methodology for evaluations to a certain extent without taking away the freedom from individual evaluators to conduct evaluations in their own style. This may involve such issues as sampling strategy and intensity (both plantations and JFMCs), formats for information collection, and practices for holding village-level meetings with JFMCs. Some evaluators, who do not have technical expertise in forestry, may benefit from exposure to forestry issues.

IMPROVED FOLLOW-UP

It often happens that an evaluator carries out investigations in a secretive manner, and there is a lack of openness in the evaluation process. This results in lack of involvement on the part of the FDA authorities in the evaluation process. In such cases, negative findings of the evaluation may not be accepted easily by the FDA authorities. On the other hand, evaluations carried out in a participatory spirit, where the FDA authorities are involved in the evaluation process from the beginning, will result in greater acceptance of the final evaluation by the FDA authorities and should result in improved follow-up on their part. It is, therefore, advisable for evaluations to be carried out in a participatory manner with the involvement of the FDA member secretary/CEO and other FDA staff. In fact, even JFMC members should be involved in the evaluation process. The method of involving all stakeholders in the evaluation process will depend on the ingenuity of the evaluators.

The drafting of the evaluation report also affects the follow-up action. Generalized comments may not result in any follow-up action. Specific observations, clear-cut recommendations, specific points for compliance, and targets, if any, should be given wherever possible in the evaluation report. This makes it easier for the FDA authorities to comply with the evaluation report and results in better follow-up.

The evaluation process should also involve the nodal office looking after FDAs and chief conservator of forests (CCF)/principal chief conservator of forests (PCCF) in charge of FDAs. They should be informed about the evaluation before it begins. The evaluators should also give a brief report to the nodal officer about the findings of evaluation, in person, through telephone, or by letter. The NAP project and FDA activities should be taken very seriously by the state governments. Such briefings by the evaluator should be received with an open spirit, which will also improve the quality of follow-up.

EVALUATIONS COMMISSIONED BY FDAs

The FDAs are allotted funds for concurrent evaluation and monitoring of their projects. These evaluations are normally to be carried out through external agencies appointed by the FDA. Since the NAEB itself is conducting mid-term and final evaluations of each NAP-funded FDA, the necessity of further evaluations at the state level is not clear. It is, therefore, suggested that the NAEB should review this component of NAP projects and decide if it is really necessary to have further evaluations at the state level. If any changes are to be made in the evaluation procedure, such as having more participatory evaluation, these principles may be incorporated in the evaluations commissioned by the NAEB itself by issuing appropriate instructions to the evaluators.

Discussions with several FDA officers have revealed that the FDAs are not very clear about the modality of appointing such agencies, the evaluation formats and procedures, and the terms of reference. The funds for this mostly lie unused with the FDAs. The NAEB should have a thorough reassessment of the whole issue of FDA-level monitoring and evaluation. If it is decided that such evaluations are to be carried out, these evaluations should add value to the project beyond what is being achieved through the NAEB-commissioned evaluations. There should be a difference in the procedure for such evaluations; otherwise, there will hardly be any difference between these evaluations and the NAEB-commissioned evaluations. The NAEB should clarify its ideas on this subject and issue guidelines to the state governments for carrying out these evaluations. If evaluations are to be carried out at the state level, an agency appointed by the FDA itself loses its independence to a certain extent. Therefore, such agencies should be appointed by a higher authority such as the PCCF.

CONCLUSION

The NAP project is a major instrument for afforestation activities in the country. Improvement in the quality of evaluations by independent evaluators is an important objective for the NAEB. In this chapter, a critical assessment of the evaluation methodology has been made, and the issues of reliability of inputs and objectivity have been discussed. Some suggestions have also been made for achieving these objectives. Confidential interviews of JFMC members have been suggested as a method for eliciting information that might not be easily available in public settings. A participatory approach to evaluation has been suggested, calling for involvement of FDA authorities as well as JFMCs in the evaluation process; this is to increase acceptability of the final evaluation and ensure better follow-up. Involvement of state-level officers, including nodal officers and particularly PCCF, sharing the results of the evaluation with them, will ensure better follow-up action.

The statements in this chapter are opinions of the author. Doubtlessly, the approach can be improved upon by inputs from NAEB officers, FDA officers, and other experienced evaluators.

CHAPTER 4

Critical Findings of NAP Implementation: An Analysis of Monitoring and Evaluation Reports

V K Sinha and H S Gupta

INTRODUCTION

Since its inception, the National Afforestation and Eco-development Board (NAEB) has been working on newer and innovative models of afforestation-related activities. Its effort has always been to evolve models of afforestation that are not only efficient and effective but also useful to the local populace. Keeping this in view, a lot of such afforestation activities have been taken up in different states of India under its umbrella programme called the National Afforestation Programme (NAP). It is quite natural for the NAEB to assess its programme and see how it is performing. The local implementing agencies, forest development agencies (FDAs), and state forest departments (SFDs) have also been conducting their own monitoring and evaluation (M&E) exercise for this purpose, besides independent evaluation being commissioned by the NAEB in selected FDAs. Further, the NAEB has been instrumental in providing a forum to independent evaluators, related FDA functionaries [through their chairperson and chief executive officers (CEOs)], nodal officer of SFDs, representatives of the government, representatives of the Forest Survey of India (FSI), academicians, and practitioners who have anything to do with the M&E of NAP to deliberate and share their findings and experiences in different workshops. The NAEB conducted a series of seven such workshops in Solan, Mumbai, Shillong, Bhopal, Bengaluru, Jadavpur, and Delhi in 2005. The various comments, ideas, opinions, suggestions, and recommendations that emerged from these workshops (thus representing a very wide cross section of stakeholders involved in the M&E of NAP) have proved to be a great resource. The current chapter attempts to analyse this expressed wisdom by classifying them into various major heads and related subtopics. In this chapter, all the views, ideas, and suggestions related to the M&E aspect of the NAP have been presented in an analytical manner so that all stakeholders (be it planner, implementer, or evaluator of the NAP) can act in an appropriate direction to achieve the stated objectives of the NAP.

ENTRY-POINT ACTIVITIES

Community participation is a vital ingredient for forestry programmes to achieve the desired goals. That is where entry-point activities (EPAs) step in, since they can be the levers that mobilize the community. In the workshops, everybody seemed to be unanimous on the fact that EPAs should be increased and strengthened, and that there should be increased funding for EPAs. Both the officials and the communities felt that the current 15% of the plantation cost earmarked for EPAs was inadequate.

It was felt that EPAs must be selected in a manner whereby they would build assets and generate some income for the participants. Moreover, communication between local communities and the concerned forest department should be more effective so that people's aspirations could be met and appropriate EPAs could be planned. A vital point raised was that EPAs should include not only ecological but also socio-economic factors.

It was opined that tardiness in undertaking EPAs was harmful and wherever such activity was started, it should be carried out within the set time frame. EPAs must not be limited only to the entry stage; it should be continued in the entire duration of the project.

COST NORMS

Several suggestions pertaining to the cost norm of the NAP that emanated from deliberations are as follows.

- The FDAs should be allowed an increased budget, necessitated by the increasing daily wage rates. It was further proposed that wage rate revision and rate of inflation (escalation of at least 10% per year) be built within the FDA-level project approval. The power to fix escalation rate should be vested with the state nodal officer.
- While deciding cost norms, geographical and topographical factors should be taken into consideration. For instance, in Nagaland it is essential to

revise the cost norms for weeding, as more funds are needed due to heavy rainfall and resultant quick growth of weeds. In such areas, there should be an increased outlay for maintenance of plantations. Transportation costs over the hills are also high; hence, allocation of funds under this overhead should also be increased.

- The amount allocated for fencing expense might be raised from the current 5%–10%, keeping in view the need for barbed wire fencing to better protect plantations.
- The NAEB should provide overhead expenditure for maintaining financial records of FDAs, as creation of a separate cell is necessary for this purpose in every FDA.
- A provision of rewards and benefits should be made for village forest committees (VFCs) that have been performing well, since this can act as a good incentive.
- The FDAs should be permitted to buy vehicles out of the overhead money, although the guidelines suggest that infrastructure is state government's responsibility.
- Agroforestry and farm forestry models should also be included as plantation models under the NAP.
- The FSI suggested that FDAs must strictly adhere to the density of plants to be raised under all plantation models to enable proper monitoring.
- The need for adopting improved and modern technology was strongly felt. For this, increased funding was suggested since this required heavy investment.
- Pro rata provision for soil and moisture conservation works, capacity building of field-level staff, better record keeping, and setting of monitoring systems at grass-roots levels should be enhanced by the NAEB.
- Pro rata provision for soil and moisture conservation work should be doubled since the current allocation of 1% is inadequate.

RELEASE OF FUNDS

The overriding concern of all participants was the constraints in the fund-release process. It was generally felt that delayed release of funds at source (that is at the NAEB) hampered microplanning. Even earlier, it did harm in implementation. The common point that emerged in the discussions was that the flow of funds was irregular and, in most cases, delayed. In some cases, the joint forest management committee (JFMC) members were not aware of the funds received under FDAs. Since the NAP was the outcome of a merger of four different afforestation schemes, the merger being done primarily to ensure timely flow of funds, the erratic funds flow was a serious concern. Also due to the delayed availability of funds, payments were kept pending for a long time. In some cases, as in Debgarh in Odisha, it had led to delay in plantations. Hence, the consensus was that there should be a definite and more efficient methodology for funds release as well as a set time frame.

To do away with delayed payments, it was suggested that efforts must be made to have electronic clearance of funds by banks. This has become easier with the introduction of core-banking system in the country. With this, the practice of using demand drafts can be done away with and instead cheques that can be cashed at any bank branch across the country could be used. This will simplify as well as quicken the procedure.

Another consensus that emerged was that since plantations were time-bound activities, the Ministry of Environment and Forests (MoEF) should be slightly flexible. It can include the option of payment in installments to the FDAs. A useful suggestion that came up was that the funds for the first year of the plantation could be allocated at one go, and the next installment could be released after 60% of the previous funds were utilized. On its part, the NAEB assured that a regional workshop on reports and documents to be furnished for the timely release of grants would be organized for chairpersons and chief executive officers of FDAs. A committee with representation from various states was formed to look into the delay in the release of funds. Further, electronic-clearing service should be considered actively for direct transfer of funds to FDAs.

PROJECT FUNDING

Since funding is a definite constraint, it was suggested that funding from alternative sources should be explored for JFMCs. Moreover, JFMCs need to be made financially more independent. It was felt that village development funds (VDFs) should be formed by the FDAs for the sustainability of JFMCs. The VDFs need not be limited to voluntary contributions from villagers. Earnings from non-timber forest products could also be deposited in the VDFs and could be used for the maintenance of EPAs, for skill development of JFMC members, and for giving loans to needy villagers and self-help groups. Utilization of VDFs must be left to the discretion of villagers.

The survival percentages of plantations indicate that the scheme has been very successful. A proposal has been sent to the Planning Commission to allocate an additional amount of ₹1150 million for the FDA mechanism. Various other ministries, such as that of textiles and tribal welfare, have also been requesting implementation of their schemes through FDAs, thus enlarging the scope for more funds. In this scenario, the role of FDAs assumes greater importance.

In addition to more funding, another factor that was considered vital was transparency in the utilization of funds. To ensure this, public boards could be put up mentioning the allocation and utilization of funds. This would help in maintaining strict fiscal discipline since the actual expenditure incurred in the field has to be reported to the parliament. Some important suggestions regarding increased funding include the following.

- The FDAs should be linked to livelihood support systems and income-generating programmes.
- Cost escalation should be strictly avoided.
- The NAEB could devise a revised funding model based on the shortfall of funds.
- In the cases of sufficient funding to FDAs, the targets could be revised.

IMPLEMENTING AGENCY

There was a unanimous view that synergy should be developed with other line departments, lateral funds should be tapped by the FDAs, and the line departments should be involved by the FDAs at the very early stages such as EPAs and other village development activities.

As of now, the relationship between VFCs and FDAs is authoritarian and one way; it needs to be made more participatory. Efforts should also be taken to make executive bodies more effective and interested in the ongoing activities. There should be regular meetings wherein the participation of executive members should be ensured.

It was also felt that record keeping of the various aspects of the FDA project implementation should be improved. Thorough documentation of all financial records, appropriate binding, and proper formatting of the record book could be immensely helpful in keeping the financial records updated. Another important point that emerged during the discussions was that frequent transfer of forest officials adversely affected the continuity of the programmes. Some check is needed to prevent this. Moreover, inadequate field staff in the forest department is a major handicap in the successful implementation of the programme. This problem needs to be addressed post-haste.

MONITORING AND EVALUATION

For any programme to succeed, constant monitoring and evaluation are essential, as these have a direct bearing on the final outcome. Thus, it was emphasized that competent evaluators should be appointed under whose supervision the entire evaluation exercise would be carried out. The MoEF should take special care of the quality of evaluation, and the evaluators must be told to be specific and maintain clarity in their findings. Under no circumstances should the evaluators generalize issues and make sweeping statements. To ensure this, the parameters and indicators for the evaluators should be objectively preset to avoid any confusion. However, some deviation from the mean seems inevitable, especially considering geographical specifics such as hilly terrain. Ideally, plantations need to be monitored and evaluated after allowing two years of growth. In hilly areas, it was noticed that although schemes for the FDAs were sanctioned as long back as 2001/02, plantations could be carried out only as late as 2003/04. In such cases, there was a compulsion to carry out the evaluation process in the first year of plantation, thus making room for inappropriate evaluation. Also, in many cases, plantations were not according to the standards laid down in the NAP guidelines.

A mechanism for direct fiscal transfer from the Government of India to FDAs entails responsibilities requiring frequent and rigorous evaluation. Therefore, it is imperative that the evaluator is fit and capable. Further, the project manager should also take responsibility of the evaluation findings and make the needed improvement. Before setting out, the evaluators should be given clear and proper guidelines so that they can communicate with all stakeholders through proper channels. Evaluators should also get in touch with the nodal office to get a perspective of the state involved. Thus, the evaluators should be totally impartial and objective in their approach and should have more interaction with the ministry as well as with the ground-level stakeholders. If the evaluation is completely from top to down, ownership of the evaluation cannot be expected at the bottom level. While the evaluation is still on, various stakeholders could be involved in the process from the beginning. This will help ownership issues. In the workshops, a grading scheme for the evaluators was suggested. The evaluators should meet the concerned officials before starting the evaluation. After visiting the field, they should apprise the CEO of the weaknesses or shortcomings observed by them. Some of the findings of the evaluators could be extremely crucial at that stage.

The evaluators should direct their energies in assessing whether the project objectives have been met or not. This is possible if the evaluators are experienced and have a proper understanding of the background of the project. Also an evaluator well versed with the job at hand could offer valuable corrective suggestions.

A need was felt to standardize the methodology of assessment and evaluation. The current methodology needs improvement. Regional concerns should be woven into the national scenario. Evaluation could be long term or short term; hence, it is necessary to chalk out the time limit for evaluation accordingly.

The criteria for selection of sites to be evaluated should be objective. If evaluation is done only for those sites that are proposed by the forest department, the results could be biased. It is important to fix an appropriate sample size to cover the social and physical aspects of the project. Uniformity in sampling methods should be maintained, particularly in those cases where there are more variations in results. Evaluators generally base their findings on 5%–10% intensity of sampling, depending on the cost and time provided for the work. Obviously, more intensified sampling would provide more authentic results.

Emphasis should be laid on self-monitoring and self-assessment. For this, a well-defined internal monitoring system should be developed. It was felt that there was a need to adopt modern tools for M&E. These could include the programme evaluation and review technique and the critical path method. The modern tools would help obtain a realistic picture; though on the flip side, these tools could prove to be a bit expensive.

Suggestions for reviewing the current evaluation format were also received from some quarters. It was proposed that the review must consider the following parameters—project area, project information, FDA mechanism, project management, people's participation, forestry and technical issues, and capacity building. The MoEF should establish the parameters; the standards and benchmarks acceptable to the ministry must be clearly defined and communicated at every level.

The participants also stressed on the proper documentation of the evaluation proceedings. It was felt that this would ensure faster and more accurate evaluation. Local leadership plays an important role in a project's success and in achieving sustainability. Thus, the evaluation pro forma must also include the role of local leadership in the promotion of local linkages. People's participation can also be gauged better if some set indicators for measurement are there, rather than depending solely on the judgement of the evaluator.

MICROPLANNING

It was felt in the workshops that microplanning must be locale specific and not simply a mindless replication of other plans. It needs to be easily understood by all the participants.

The FDAs should strive for alternative income-generation activities for the involved villagers. Income-generation options for individual members of the JFMCs should also be thought of. Regular reviews and upgrade of the microplan are also needed.

PROJECT DURATION AND MAINTENANCE OF PLANTATION

The FDAs should lay stress on maintenance and support for the initial period, say five years. Measures should be evolved to protect and maintain the assets created under the project. The thrust should be on better maintenance of plantations to improve the survival rate.

Record keeping with regard to plantations also needs to be improved. Plantation register and logs must be maintained painstakingly so that the statistics needed is immediately available. The stock registers as well as other relevant financial records should be properly maintained.

OBJECTIVE OF SCHEME

Inclusion of all village communities is important for poverty alleviation and sustainability. Certain communities are always excluded; it is often the dominant groups that are included in the scheme of things. The FDAs should take care to see that all sections of the society are included in the programme.

Plantations should be done in such a manner that they ensure maximum benefits on short-, medium-, and long-term basis. The general perception was that employment generation from the project was much less than anticipated and efforts should be made to generate jobs from it throughout the year, rather than seasonally as has been the case so far. Equity, whether of gender or any other social aspect, should also be considered seriously. The FDAs can also be the nodal agency for the development or creation of a database of the local species and local practices used for conservation. Along with income generation, market linkages for all activities should be developed. The FDAs can also be a platform for marketing channels. Yet, it was remarked that the primary objective of the FDAs was to conserve and maintain the cover of high density forests; rural development was not the primary objective of the forest department. It was felt that to ensure more public participation, FDAs should curb the tendency to employ outside workers. Efforts should be made to have an all-inclusive participation for which regular meetings should be held. All villagers should be included in the committees subject to their willingness. All strata of the society should be represented.

PROJECT AREA

Here the common denominators were that plantations had not been done according to the NAP guidelines. In many cases, it was found that plantations were clustered at one place rather than equitably distributed in the area. In many cases, field parties from the FSI observed that plantations were done in a single patch where place was available, instead of spreading them over the whole area.

It was felt that the NAP should not be implemented in an area where other joint forest management (JFM) projects were being implemented, as that would create confusion.

PROJECT PLANNING

On planning of projects, the most oft-received suggestion was that more groundwork was required before launching the project. For instance, base data need to be collected in the preparatory stage itself, before the actual implementation starts. A standard manual needs to be compiled for the FDAs to perform their activities. The need was also felt for forecasting and monitoring the benefits of a project before launching it.

It was realized that entrepreneurial development programmes were essential for the sustainability of JFMCs. The formation of self-help groups was also emphasized. It was suggested that in the remaining years of the Tenth and Eleventh Five-year Plans, more emphasis should be laid on strengthening JFM to increase people's participation and provide sustainability. However, this gave rise to the bigger question as to how that would be achieved when project funding had been stopped.

Along with community participation in projects in general, suggestions also came for increased participation of women. This could happen with the formation of more women-centred self-help groups.

Another vital suggestion was for selecting species in the plantations that conformed to the agro-practices of the local populace. For instance, in the state of Punjab, people did not plant eucalyptus trees extensively. It was being planted on roadsides and waterlogged and paddy cultivated areas. Therefore, the recommendations were in favour of planting native and fast-growing species. There should be better planning by selecting mixed species for the plantation on the basis of short-, medium-, and long-term rotation period.

However, by far the most relevant suggestion was setting realistic targets. Very high targets for JFMCs need to be given up. Non-governmental organizations need to be involved in the process.

IMPROVED TECHNOLOGIES AND TREATMENT OF PROBLEM LAND

It was overwhelmingly felt that modern technology should be encouraged in the plantations at the local level. Improved technology—including use of quality seeds and plantation material, root trainers, rhizobium inoculation, mycorrhizal inoculation, hormonal treatment, bio-fertilizer, and integrated pest management—should be made an integral component of approval of FDA projects.

CAPACITY BUILDING

Improved communication was the key suggestion in this area. It was felt that there was a communication gap between planners and the projected audience (villagers). It was felt that the concept of JFM was present merely on paper. The communication between the local populace and the FDAs was practically low, and this needed to be rectified. For the successful implementation of the programme, effective communication could prove to be a very effective tool. To reach out to the ground-level staff and form effective communication channels, training schedules should be organized for the field-level staff. They must be given proper briefing of the scheme guidelines and its aims and objectives. One type of training required is for financial management and proper documentation and maintenance of records.

Training areas should be identified further and well-structured training should be imparted.

Capacity building can thus be achievable by effective field-based training for forest staff. It was, therefore, suggested that the focus be on capacity building of the community. It was also suggested to train village forest management and protection committees on project objectives, strategy, sustainability, components and benefits, choice of species, income generation, integration with village development, convergence, and skill transfer.

CONCLUSION

The findings of the deliberations that took place in various regional workshops have been analysed in this chapter. These are quite valuable from the planners' perspective as they give the planners enough ideas of the strengths and pitfalls of the schemes, institutions involved, and which aspect needs to be modified while improving the current scheme of the NAP. A study of this analysis will also help the implementing agencies (that is, FDAs) in better translation of the ambitious greening programme of the NAP. The evaluators (both departmental and independent) can equally benefit from this analysis, which offers them nuances of M&E in integrated perspectives. These are also useful to implementers as well as SFDs. In fact, all experts who deal with afforestation-related work (in particular) and anyone working in the field of M&E will find this analytical study quite useful in the Indian context.

CHAPTER 5

Monitoring and Evaluation Issues in the Watershed Development Component of Natural Resource Management

N Sahu and G B Reddy

INTRODUCTION

Watershed development supports the livelihoods of the poorest of the poor in watershed areas. After careful evaluation of watershed management projects rooted in the classical soil conservation technology, the following can be said about watershed management (WSM) practised in these projects. WSM has been adopted as a strategy for natural resource management in India by government and non-government agencies for the last 10 years. While some remarkable progress has been achieved in some areas, the overall impact is not very encouraging. Farmers have often refused to adopt the recommended practices for a variety of reasons or have abandoned them once the project support was withdrawn. Soil and water conservation structures frequently collapsed due to poor construction, lack of maintenance, and sabotage.

In the last one decade, a lot of innovations have happened that transformed watershed development from a classical soil conservation approach to a sustainable livelihoods approach. Watershed development now operates in a participatory platform where the community plans, implements, and monitors the programme.

Participatory monitoring and evaluation (M&E) has been an integral component of programme design. Thus, designing an M&E system for any watershed and natural resource management programme is necessary as it helps in providing a constant guidance for the project performance at all stages. Monitoring and evaluation also facilitate in providing project learning.

Monitoring involves tracking the progress of an act on a continuous basis. It can be defined as a function that supervises the ongoing process of intervention to provide the project staff, programme managers, and key stakeholders with indications about the progress, or lack of it, in achieving the targets. Thus it does so by measuring inputs and outputs and any change in the outputs due to change in the inputs.

OBJECTIVES OF AN M&E SYSTEM

Broadly, the objectives of an M&E system are as follows.

- Providing feedback on performance of the project.
- Helping meet accountability and transparency standards.
- Providing inputs for planning, strategizing, and decision-making.
- Contributing to project learning and advocacy.

GUIDING PRINCIPLES OF AN M&E SYSTEM

The M&E system and its development process are influenced by various factors such as the background of M&E, institutional priorities, expectations from M&E, skills, level of operation, and efficiency, which vary from stakeholder to stakeholder. For example, to satisfy the larger requirements of the government, the emphasis is on physical and financial progress reporting. Monitoring and evaluation are normally used to identify gaps and leakages in quantitative progress and take corrective measures. Monitoring of the process and qualitative learning are rarely attempted. Hence, the nature of a contemporary system requires focus on the quality and process aspects of monitoring, in addition to quantity. This interplay of multiple factors, which are conflicting at times, necessitated the laying down of M&E principles, which mediate between contrasting demands of the stakeholders from the system and ensure a commonality of understanding, purpose, and vision.

To comprehend and understand a contemporary system, the following guiding principles may be laid out.

- The system should be developed and operated according to the demands of M&E from various stakeholders.
- The system should be simple and easy to operate, which will help functionaries in building their skills and leading the execution process.

- The system should be participatory. It should be developed with the joint initiative of communities and staff, which will increase the ownership of the system, and it should not be used for policing and punishment.
- The system should encourage reporting mistakes and errors as part of a learning systems approach. It should not police but should encourage people to take initiative and learn.
- The system should promote an appropriate mix of both quality and quantity and ensure balance.
- The system should cater to both regular and one-off monitoring requirements. While building a regular system, this satisfies the one-time information requirements of some stakeholders.
- The system should contribute to "small learning loops". Continuing with an earlier principle, this encourages learning and dissemination of learning at appropriate levels.
- The system should facilitate appropriate aggregations rather than sending all data upwards. This reduces load of data at the higher management level.
- The system should function as a feedback system and enable two-way passing of information. This not only ensures generation of information but also gives feedback to respective levels for actions.
- The system should emphasize on both data collection and analysis.
- The system should also give emphasis to data analysis and use at source as well as eliminate ignorance and appropriate imprecision.

WHAT TO MONITOR?

A system can monitor several things, which would be time consuming. Therefore, it needs to be strategic in choosing what to monitor, how to monitor, and how much to monitor. Some of the broad areas that an M&E system can address include the following.

- **Inputs:** Money, material, and human resources.
- **Activities:** A set of tasks performed such as self-help group meeting held and plan for WSM prepared.
- **Process:** A set of activities towards a common purpose such as mobilization of community contribution and building of WSM.
- **Outcomes:** Series of effects of some action and activities such as reduced run-off, increased recharging, and increased water table.
- **Impact**: Changes in peoples' lives and livelihoods such as increased income, reduced morbidity, and increased capacity to negotiate.

- **Outputs**: Project logical framework (LOG FRAME) defined such as the livelihood asset base for the poorest is enhanced and diversified.

MONITORING AND EVALUATION FRAMEWORK

The M&E framework consists of tools and mechanisms for monitoring related to inputs and activities, processes, outputs and outcomes, and impacts and capacities. Evaluation involves impact evaluation, programme evaluation, and institutional evaluation. The M&E framework of the Western Orissa Rural Livelihoods Project (WORLP) is given in Figure 5.1.

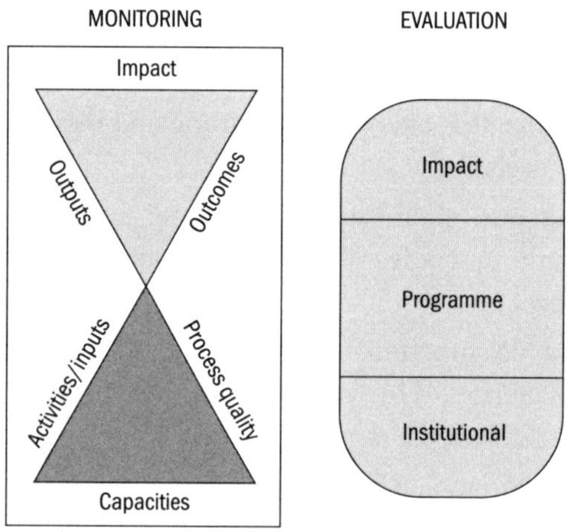

Figure 5.1 *Monitoring and evaluation framework of WORLP*

LEVELS OF MONITORING

Levels of monitoring are identified according to the level of stakeholders' involvement. For example, the M&E system of WORLP operates through the following four levels, which are important for management, data aggregation, learning, and feedback.

1. **Watershed level:** Watershed committee, users groups, and self-help groups.
2. **Block level:** Project implementing agency and livelihoods support team.
3. **District level:** Project director and capacity-building team.
4. **State level:** Directorate, Orissa Watershed Development Mission, and project support unit.

TOOLS AND MECHANISM

Table 5.1 gives a complete list of the tools and mechanisms used in the M&E system and their description against each identified area for monitoring.

INDICATORS FOR M&E

Indicators provide the main criteria for M&E that specify performance standards, target groups (for whom), quantity (how much), quality (how well), time (by when), and location (where). Measurement against each indicator at different points of time reflects the changes that have taken place during that period. An example of a project LOG FRAME with objectively verifiable indicators (practised in WORLP) is appended in Annex 5.1.

INSTITUTIONAL FRAMEWORK

There needs to be an institutional and financial provision to make the M&E system of any project effective and operational. Monitoring core groups are the mainstay of the WORLP M&E system; development and implementation of the system takes place through these groups (Figure 5.2). At the state level, the state monitoring core group provides strategic direction to the M&E system and guides the implementation process.

EVALUATION

Periodic evaluations are mandatory in any project. The evaluations are conducted by internal or external agencies or individual or both. Evaluations assess the project performance in terms of outputs, outcomes, processes, and impacts.

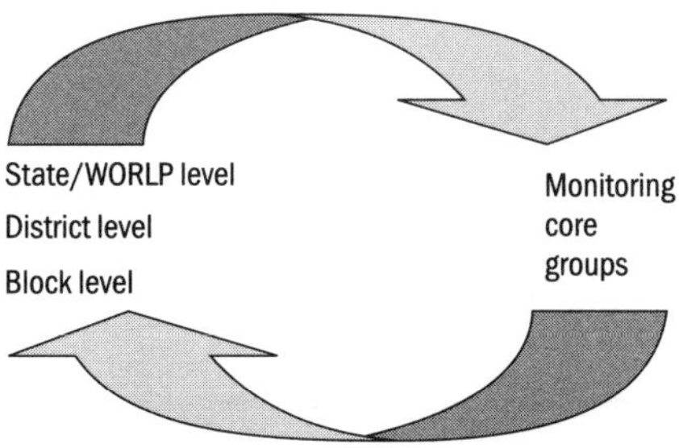

Figure 5.2 *Framework of WORLP M&E system*

Table 5.1 Tools and mechanism of an M&E system

Area of monitoring	*Tool/mechanism*	*Description*	*Level of application*
Input/ activity	Activity report (quarterly, end of six months, annual)	Physical progress report of inputs and activities	All levels
	Simplified microcredit accounting system	Progress of microcredit activities of self-help groups	Watershed level with self-help groups
	Financial assurance system	Financial monitoring related to financial systems, procedures, tools, and templates	All levels
Process	Peer review system	A review mechanism by which the project director, capacity-building team, area project director, project implementing agency, livelihoods support team, watershed development team of one district or block review the work done by their counterparts of another district or block	Watershed level with self-help groups, user groups, and watershed committees
	Process monitoring	Qualitative process indicators developed to monitor implementation processes	All levels
	Participation index	Community participation in project activities	Watershed level with watershed association, self-help groups, user groups
Output/ outcome	Output to purpose review	Contribution of outputs towards the achievement of the purpose, impact	All levels
Impact	Measure of livelihoods	Measurement of change in livelihoods asset position	Watershed level with self-help groups or user groups
Capacities	Group self-assessment	Self-assessment of the strengths of self-help group or user group or watershed committee	Watershed level with self-help groups, user groups, and watershed committees

On the basis of basic information, reports, and field observations, assessment or evaluation is carried out to quantify or rank the performance of the projects against pre-stated performance indicators. The evaluation findings strongly influence decision-making processes such as release of funds, continuity of project, including mid-course corrections.

Sometimes, the project evaluation framework is predefined and is a part of the guidelines, whereas at other times the methodology and indicators are worked out with and agreed by stakeholders. A predefined evaluation framework for forest development agency is defined in Annex 5.2.

GAINS FROM AN M&E SYSTEM

A robust M&E system establishes a baseline to create benchmarks on which the performance of a programme can be evaluated. It checks the pulse of the project and informs stakeholders to take appropriate measures at different stages of the project cycle. The M&E system also informs project participants to what extent the project has achieved its objectives. The system helps in creating opportunities to review the plan and encourage mid-course corrections. It contributes towards the learning loop of the project so that the best practices are documented and widely replicated. The inclusion of participatory monitoring has made best use of the information at source. Thus, the community has set examples in terms of empowerment, transparency, and ownership of the project.

ANNEX 5.1

Project LOG FRAME (WORLP)

Narrative summary	*Objectively verifiable indicators*	*Means of verification*	*Assumptions for success*
Goal: Adopting more effective approaches to sustainable rural livelihoods by government agencies and other stakeholders in the KBK districts and elsewhere	• Population below official poverty line fell by 30% by 2010 in project districts. • New government guidelines and schemes that improve participation, equity, and convergence were adopted by GoO by PY5. • Donors prepared to move from project to sector support during the second half of the project.	• Census 2011 • MoRD/GoO guidelines • Government and donor reports	
Purpose: Promotion of sustainable livelihoods, particularly for the poorest, in four districts in replicable ways by 2010	• At least 70% of the poorest men and women in the project villages reported significant increase in livelihood assets by EOP. • At least 70% of marginal farmers reported improved disaster-coping capacity as a result of project interventions by EOP. • Exploitation of migrants and the incidence of distress migration were reduced. • Enhanced access to and control of natural resources were reported by the poorest by EOP. • Returns from land, water, and livestock were sustainably increased through improved management. • Child malnutrition was reduced by 30% in project areas.	• Baseline survey • Impact assessment studies • EOP evaluation • Special studies • Participatory M&E reports • Process documentation • Reports from other agencies • Disease surveillance system	• Other development schemes of GoI and GoO having a direct bearing on the livelihoods of people were effectively and successfully implemented. • GoI and GoO policies were supportive, and resources were made available for replication and associated capacity building.

Contd...

ANNEX 5.1 *Contd...*

Narrative summary	*Objectively verifiable indicators*	*Means of verification*	*Assumptions for success*
	• Morbidity was reduced by 30% on account of decrease in malaria and waterborne diseases in project areas. • Improved access for the poorest to natural resources, government institutions, and services was enabled by an improved policy environment. • CBOs negotiated regularly and effectively for rights and entitlements with government, PRIs, and NGOs.		
Output 1: The poorest were organized and able to effectively plan and implement participatory livelihood focused development.	• At least 75% of the poorest households were part of the viable and active CIG/SHG by the third MWS year. • At least 30% of WDC members in each MWS were women who reported active involvement in all phases and activities of the project by EOP. • At least 70% of the poorest reported inclusion of key interests in MWS microplans at MTR. • In all WORLP MWS, microplans were developed, implemented and regularly reviewed in a participatory manner by PY7. • At least 50% of village specialists trained by the project come from the poorest households and at least 30% were women by EOP. • At least 60% of WDCs operated and effectively managed project-created assets beyond MWS year 5. • At least 50% of the poorest in the project area reported increased access to and benefited from common property resources, project assets, and amenities by EOP. • The poorest were able to act as pressure groups to access services through their own initiative by EOP.	• Evaluation reports • Village-level records and documents • Participatory M&E reports • Special studies • Capacity-building plan documents • Community-level documents • Project progress reports	• Participatory planning processes were recognized and supported by district administration and GoO. • PIAs accepted and effectively implemented project approaches and practices. • Village communities accepted meaningful participation of their poorest members in decision-making.

Contd...

ANNEX 5.1 *Contd...*

Narrative summary	*Objectively verifiable indicators*	*Means of verification*	*Assumptions for success*
Output 2: The livelihood asset base for the poorest was enhanced and diversified in 290 watersheds.	• Access by the poorest to financial services and credit was improved. • At least 70% of the poorest reported an increase in the months of employment and enhanced income generation activities by EOP • At least 35% of the poorest households adopted improved sanitary practices by PY7. • At least 35% of women reported reduction in agriculture- and household-related drudgery by PY7. • Incidence of lean season food shortages and malnutrition reduced in 290 watersheds by EOP. • Returns from rain-fed land increased by 30% by PY7. • Returns from livestock/aquaculture, particularly for the poorest, increased by 30% by EOP. • At least 90% households have access to permanent safe drinking water within 100 m of house. • Incidence of malaria reduced by 10% through adoption of preventative practices.	• Impact assessment studies • Project evaluation reports • Special studies • Participatory M&E reports • Partner agencies reports • Project progress reports	• Successful livelihood activities were replicated elsewhere in KBK. • GoO promoted project approaches in KBK districts.
Output 3: Government, PRIs, and NGOs together effectively implemented participatory livelihood focused watershed development.	• Performance criteria for PIAs, other implementing agencies, and OWDM were agreed by the end of 2004 and implemented in accordance with agreed milestones. • At least 50% of the poorest reported increased access to relevant information and government resources at MTR and 100% at EOP. • At least 50% of PRI representatives in project villages reported active participation in project activities by EOP.	• OWDM-level policy documents and guidelines • Performance review reports • Non-WORLP project reports • MWS plans • Project progress reports	• Secondary stakeholders and communities agreed to improve their way of working together on pro-poor issues. • Government agencies promoted participatory processes and approaches.

Contd...

ANNEX 5.1 *Contd...*

Narrative summary	*Objectively verifiable indicators*	*Means of verification*	*Assumptions for success*
Output 4: Policy and practice constraints to livelihoods of the poorest were reduced in the areas of NTFPs, migration, land rights, disaster preparedness, and women's issues.	• Constraints to pro-poor policies and practices were identified by PY5. • Pro-poor changes were made to government policies and practices by PY7.	• Special studies • Progress reports • Government reports • Workshop proceedings • Process documents	• Changes to policies and practices were accepted and implemented by government agencies.
Output 5: Project approaches were adopted in KBK and elsewhere in Odisha.	• Project approaches and interventions were promoted and adopted in 50% of non-WORLP MWS in the four project districts by EOP. • Participatory livelihood focused watershed development best practices and processes developed by WORLP were replicated in most watersheds in KBK districts and elsewhere in Odisha by EOP.	• Non-WORLP MWS plans • Non-WORLP project reports • Project literatures, newsletter, articles, websites • Feedback from workshops and conferences • Project progress reports	• OWDM issued guidelines and directives supporting adoption of project approaches elsewhere. • Government and other agencies promoted project recommendations widely.

CBO – community-based organization; CIG – community interest group; EOP – end of project; GoI – Government of India; GoO – Government of Odisha; KBK – Kalahandi–Balangir–Koraput; M&E – monitoring and evaluation; MoRD – Ministry of Rural Development; MTR – mid-term review; MWS – micro-watershed; NGO – non-governmental organization; NTFP – non-timber forest product; OWDM – Orissa Watershed Development Mission; PIA – project implementing agency; PRI – Panchayati Raj institutions; PY – project year; SHG – self-help group; WDC – watershed development committee; WORLP – Western Orissa Rural Livelihoods Project

ANNEX 5.2

EVALUATION FORMAT: FOREST DEVELOPMENT AGENCY

Grading of project on a scale of 1 to 10

1. **Quantitative aspects:**
   - Physical achievements
   - Financial achievements
2. **Qualitative aspects:**
   - Survival of plantation
   - Health of plantation
   - Maintenance of plantation
   - Sustainability of plantation

Grading of project on a scale of 1 to 10

3. **Joint forest management:**
   - Degree of people's participation
   - Satisfaction over perceived benefits by the local population
4. **FDA mechanism:**
   - Composition of general and executive bodies of the FDA
   - Role of FDA in the administrative, supervisory, and monitoring mechanisms of the project
   - Flow of funds from FDA to VFCs
   - Planning, implementation, and maintenance of assets directly by village forest committees (VFCs)/eco-development committees (EDCs)
5. **Overall grading of project on a scale of 1 to 10:**
   - Outstanding (8–10)
   - Very good (5–8)
   - Good (3–5)
   - Poor (<3)

CHAPTER 6

Developing Effective Sampling Techniques for Monitoring and Evaluation of FDA Projects under NAP

VK Sinha and
CVRS Vijay Kumar

INTRODUCTION

The National Afforestation Programme (NAP) is a 100% centrally sponsored scheme in the Tenth and Eleventh Five-year Plans of the Ministry of Environment and Forests (MoEF), Government of India. The NAP institutionalizes the participatory approach to sustainable development of forests ensuring uniform funding and implementation mechanism, avoiding delays in the availability of funds at the field level and ensuring effective peoples' participation in project formulation and its implementation. The scheme is operated by the National Afforestation and Eco-development Board (NAEB) of the MoEF. The funds are provided directly to the forest development agencies (FDAs) in the forest divisions. At the village level, the scheme is implemented through the joint forest management committees (JFMCs) under different nomenclatures such as the forest protection committee, village forest committee, or village samrakshan samiti. In protected areas, the eco-development committee (EDC) serves as the JFMC. State-level and national steering committees also monitor the implementation of NAP projects.

The NAEB had issued detailed and comprehensive operational guidelines on how to seek funding support for afforestation schemes under NAP (NAEB 2005). The guidelines provide for budget allocation for different components of forest and village development activities. In addition to the monitoring and evaluation (M&E) carried out by supervisory forest officers, 2% of budget costs can be earmarked for concomitant M&E. The FDAs usually engage independent agencies such as research institutes, retired experts, and non-governmental organizations (NGOs) having expertise to carry out concurrent M&E of the NAP and FDA activities. In addition, the NAEB also gets the evaluation of NAP projects done by independent agencies/consultants. The first evaluation is done in the second year and the final evaluation in the fourth year after the project is sanctioned. This M&E work is intended to ascertain whether people's participation, functioning of JFMC/EDC, and microplanning are adequate. The Forest Survey of India also monitors the area estimates and key biophysical parameters of plantations through its national-level sampling carried out in the field.

The evaluators obtain details of FDA activities from the forest divisions and document their assessment/evaluation during the fieldwork and provide information in the prescribed evaluation format. The NAEB organizes annual regional workshops for the evaluators, FDA chairpersons (conservators of forests), chief executive officers (divisional forest officers), and state-level senior forest officers to discuss the evaluation reports and improve the functioning of FDAs, NAP project design, NAP processes, and M&E processes.

The NAEB had circulated a detailed format in 2003 for evaluating projects. This format examines the quantitative and qualitative aspects of plantations, people's participation under joint forest management (JFM), and FDA mechanism using 12 parameters (Annex 6.1). The same format is also used for concomitant M&E activities at FDA and state levels as well as by the evaluators appointed by the NAEB. The NAP has both biophysical and social aspects, which are critical for the programme implementation. Hence, evaluators are required to have both the skills to complete the evaluation format in an unbiased manner.

Although the format has served the cause of NAP objectives in an excellent manner, there is a feeling that the format is too long for the purpose of the first concurrent evaluation. There is a need to revise the format to highlight the key points. Further, the current practice of considering only a few sample plots in plantations also needs to be reviewed to reduce the bias and bring consistency in the rapid assessment approach. Focused guidelines are also required to direct evaluators having varied backgrounds, expertise, and experiences to ensure comparison at different levels.

OBJECTIVE

The objective of the research was to review the sampling design and evaluation format for NAP and the activities of FDAs.

METHODS

The study of the quantitative aspects of plantations was carried out by visiting young plantations under the NAP and other schemes. The feasibility of carrying out different sampling approaches for rapid assessment was tested in the field by laying down plots and carrying out the exercises in the field in Vidisha, Betul, Mandla, and Nagpur areas. Some of the plantations visited during the exercise were (1) miscellaneous species plantation of 2002 in Gambhiriya P 55 at Gyaraspur under fodder development; (2) teak, aonla, jamun, and mango plantations of 2003 and 2005 at Khanda in Dharampur under fodder development; (3) assisted natural regeneration plantation of 2002 in Compartment 263 of reserved forests; and (4) teak plantation of 2004 in Compartment 15 at Tilaipani in Kanchangaon and of 2004, 2005, 2006 at Nainpur.

The study of the social and implementation aspects of NAP was carried out through discussions with foresters and evaluators, and presentations in a few NAEB-funded workshops. The MoEF organized a brainstorming session on the subject, and its inputs were used to improve the format. In addition, the formats used by Madhya Pradesh and Maharashtra states were also studied.

FORMATS FOR FIELD DOCUMENTATION BY EVALUATORS AND INFORMATION FROM FOREST DEVELOPMENT AGENCIES

The components A, B, and C in the current evaluation format contain information to be furnished by the FDA concerned. A separate format may be planned for such information, which should be filled up by the FDA before the M&E field visit. Evaluators may use the information for the site sampling and attest the information in general terms based on observations and feedback collected during the fieldwork. Evaluators should document site information in a separate format based on data collected from the sample sites and sample villages along with their observations. This two-format approach will adequately segregate direct observations from the information provided by FDAs. The format for evaluators should be on key aspects. While the evaluators should submit the consolidated report in the prescribed format, the data from individual sites visited should also be appended.

SITE SAMPLING

The evaluator should select at least five to six sample sites to document quantitative information about the plantations, carry out focus group discussions and household interviews in the village, manage the sample plantations spread over the geographical area and plantation models. The rapid assessment nature of the exercise makes it necessary to restrict the number of sites. Common judgemental sampling is vulnerable to inadvertent bias due to convenience sampling. Therefore, it is necessary to define a protocol to ensure random selection of sample sites. The range of survival percentage in reforestation and afforestation areas is usually large enough to make a guess about the variance and to use the same for calculating sample size. Further, the rapid nature of assessment does not allow the sample size to be large enough to be statistically significant at the FDA level. However, unbiased and consistent sampling at the FDA level will allow subsequent aggregation of the information at regional or state level.

The information used to develop the inventory design should be related to, but distinct from, the item of interests (health of regeneration in this case), should be available for every unit, and should be inexpensive to measure (Robinson, Hamlin, and Fairweather 1999). Implementing agencies commonly document the plantations by the planting year, the plantation models, and the site along with its administrative jurisdiction by ranges. The cost implications of the fieldwork and uneven distribution of plantations in ranges do not allow proportionate sampling in each plantation model and range. The NAEB requires M&E at different intervals, and hence the planting years could be a valid parameter for stratification. Variations in site conditions are not documented in the routine consolidated reports and, therefore, cannot be used for stratification. Hence, the random number based on sample selection may be applied to a list organized in an unbiased and consistent manner (Annexes 6.6 and 6.10). The multistage sampling and cluster sampling may be problematic to many evaluators not well versed in sampling techniques.

A possible sampling mechanism could be to carry out the list sampling (Robinson, Hamlin, and Fairweather 1999) by arranging the list of all sites in the division of the given planting year (in ascending order) along with net plantation areas according to the plantation models, ranges, and compartment numbers. If a site does not have a compartment number, its revenue survey number (or name of the village) could be used for arrangement. The random number could be applied on the cumulative area. This approach will ensure that common models and large plantations will have a greater chance of being selected for fieldwork. The evaluator may add one or more randomly selected plantation(s) if it is expected to elude some significant information.

QUANTITATIVE ASSESSMENT OF SURVIVAL OF PLANTATIONS IN SAMPLE PLANTATIONS

Based on reconnaissance survey, the plantation area could be stratified using prominently topographic and biophysical characteristics such as distinct variations in slopes, soil characteristics, species, and growth conditions easily discernible in the field. However, this approach does not appear common. Perhaps, time constraint serves as a deterrent in most cases. Laying a small number of plots (even two or three plots in many cases) in a judicial manner is the most common approach. A small number of sample plots does not provide adequate information regarding the variations within the area. Moreover, the number of sample plots should vary with the size of the plantations. If this critical information is inadequate at the site level, compilation at the FDA level is likely to generate data with a large uncertainty.

Evaluation reports generally do not have documentation of the sampling design used during the fieldwork. The treatment maps given in the plantation registers are often too rough and at a scale unsuitable for planning sampling design. Representative sample plots are difficult to be laid down in small plantations.

The size of plantations under FDA varies substantially. In many cases, particularly under the assisted natural regeneration model, the net plantation area is much smaller than the gross plantation area. Similarly, many plantation sites have a number of unsuitable patches for a variety of reasons and are left blank. Although seedlings are planted according to standard alignments, different parts of a plantation may have different alignments. In some areas, plantations are taken up by subdividing the area into small sections. However, ascertaining a particular section could be time consuming because of the section boundary and its marking being unrecognizable in the field during the various plantation operations and natural processes. The plantations have a natural periodicity due to predefined spacing used at the time of alignment. Minor deviations from the stated spacing and alignment are not uncommon. Variations in shape and slope add to the difficulties.

The presence of periodicity in the arrangement of plants must be accounted for while defining the size of plots or width of strips. Two different approaches were tested in the field. Systematically laid plots of different sizes were enumerated at different places. Most of the visited sites had high survival rates; therefore, the variations due to the plot size were not discernible. The details of one site in Mandla region are given in Table 6.1 to illustrate the variations due to plot size. The plot size does not appear to have noteworthy variations in the survival count. Similarly, rectangular plots arranged in strips displayed the same range of variations.

Routine assessment of survival in plantations using line count or strip count in a systematic manner, usually with a 10% sampling intensity, is common in the forest divisions. For example, in Maharashtra, the guidelines for assessment of survival percentage include counting the strip of three rows of planted seedlings with 10% sampling intensity, that is, counting every 10th strip. In very small area of 2 ha or less, every 10th row is to be counted. Every 20th plant in the strip or row is to be measured for height and collar girth. However, there is a growing tendency to increase the sampling intensity or to carry out 100% seedling count in sample verification of plantations in many states (such as Gujarat).

Although complete count of the strips may provide a good estimation of the mean survival percentage and health of plantation, it has limited value in displaying variability in smaller plantations. Hence, it is proposed that the strip has width of spacing and multiplied by three. Documentation may be carried out in segments of double the segment width. Counting may be carried out for every third segment along the strip. Incomplete segments need not be counted. The first strip may begin at any random point, and subsequently roughly parallel strips may be repeated in such a manner, on both sides of the first strips, that the entire area is covered. If roughly parallel strips are spaced out at six times the strip width, the sampling exercise can be carried out faster with sampling intensity in the range of 5%–6%. Larger intervals up to a specified distance, say 10 times the strip width, will further reduce the intensity, reducing the cost while generating data for analysis. Counting at intervals also allows the evaluators to criss-cross the entire plantation area to make observations about the health of the plantations.

For the purpose of reporting, the survival percentage should be weighted by net area planted in the same model. The percentage should be reported separately for each plantation model covered for the counting and compared with the value reported by the FDA. If the use of the common indicator survival percentage becomes unavoidable, the survival count of individual plantations should be weighted by net plantation area and average number of seedlings in each model.

Table 6.1 Average survival percentage based on rapid assessment survey using plots of different sizes laid down in Compartment 15 Mohgaon 2004 teak plantation @ 2500 plant/ha

Items	*Plots with positions (36×36 m^2)*									
	North-east 12 × 12 m^2	*South-east 12 × 12 m^2*	*South-west 12 × 12 m^2*	*North-west 12 × 12 m^2*	*North-east 18 ×18 m^2*	*South-east 18 ×18 m^2*	*South-west 18 × 18 m^2*	*North-west 18 × 18 m^2*	*Central 24 × 24 m^2*	*Combined 36 × 36 m^2*
Area (ha)	0.0144	0.0144	0.0144	0.0144	0.0324	0.0324	0.0324	0.0324	0.0576	0.1296
Blank	7.5	9.5	8.7	7.3	16.5	16.3	14.3	20.0	30.8	62.0
Ht class <30 cm	2.0	2.0	0	3.0	2.0	7.0	1.5	4.0	2.5	5.0
31–137 cm	5.5	3.0	3.0	3.3	13.0	9.0	7.3	8.3	14.0	37.5
>137 cm	22.0	23.8	28.5	26.0	44.3	51.0	58.5	48.8	100.3	202.5
Planted	35.5	36.0	38.0	38.0	74.3	78.0	80.8	74.0	147.5	307.0
Planted/ha	2465.3	2500.0	2638.9	2638.9	2291.7	2407.4	2492.3	2284.0	2560.8	2368.8
Survival %	79%	74%	83%	81%	77%	79%	83%	80%	79%	80%

The same strip plots can be used for counting natural regeneration, coppice, and seedlings from seed sowing. It should be recorded separately. The strips should be laid down along the slope to capture the productivity gradient. The measurement of seedling height and collar girth is time consuming. Hence, it is recommended to use seedling height class as an indicator of plantation health during the first concurrent evaluation.

SOCIAL AND IMPLEMENTATION ASPECTS OF NAP

Based on the feedback of evaluators and forest officers at different levels in various NAP- and FDA-related workshops from 2002 to 2006, a draft format for the first concurrent evaluation of JFM villages was developed and presented in a series of workshops during 2007 and 2008. Feedback for improvement (Table 6.2) was obtained from different groups, and the formats were discussed with field staff and villagers during site visits. While forest officers gave greater emphasis on regeneration, forestation, and forest protection, the villagers laid more stress on employment generation, benefit flow to villagers, and village infrastructure development.

The evaluators generally felt that the current format largely meets the need of detailed feedback to the NAEB but that there is a need to further specify the monitoring parameters for participation. They also felt that the evaluation format must have enough details to clarify the current situation and help in further improvement. They were doubtful about the possibility of household interviews being feasible within the present cost norms for M&E.

The forest officers were generally in agreement with the general structure, but they recommended a higher weight to plantation works and greater objectivity in assigning grades. A few senior forest officers pointed out that FDA activities should be evaluated in consonance with the proportion of budget allocation available for different components. Since plantation operations have higher allocation in the budget, it should have a higher weight in evaluation. There will be distortion in the evaluation process if FDAs are evaluated on social parameters for which a small amount of funds is made available under NAP; FDAs have limited ability to influence these parameters. The response of other government agencies to village infrastructure development largely depends on personal rapport with individual officers of other government and non-government agencies. In spite of JFM and NAP being in operation for a long time, the required support of other agencies is still not overwhelming. Most officers are sceptical about the impact of FDA activities concerning empowerment, community values,

Table 6.2 Feedback of forest officers and evaluators on the evaluation parameters

Average score on 10 point scale	*Evaluation parameters preferred by forest officers and evaluators*
More than 7.5—most important	• Survival percentage • Natural regeneration • Quality of forests • Flow of fund and its utilization • Microplanning process • Community participation in microplan preparation and its implementation • Effective communication and transparency in the process
5.0–7.5—important	• Record maintenance • Benefit flow to villagers • Village infrastructure development • Gender equality • Village development fund • Conflict management through participation • Inclusive growth with attention to the need of disadvantaged groups • Improvement in household economy • Interaction with government agencies
2.5–5.0—less important/not relevant	• Nursery practices • Coordination with *Panchayati Raj* institutions • Use of poverty score card • FDAs' contribution to human development and people's happiness • Empowerment and political freedom • Households' relationship with others and their safety • Social cohesion and community values • Representation of disadvantageous group (see inclusive growth above)

and political freedom in the prevailing socio-economic conditions. Based on the feedback received, the draft evaluation format has been revised by excluding parameters considered less important/not relevant. The NAEB had communicated that the format should be further abridged. In the light of the feedback, it was considered inappropriate. However, the norms for objective scoring of components are given in Annex 6.4. The revised format is presented as follows.

First Concurrent Evaluation of JFMC Activities and Plantations

(Refer to the guidelines for sampling, and conducting evaluation and overall assessment)

Dates of evaluation: ________________

Period evaluated (month and year): From ______20 ______ to ______ 20 ______

1. Site details

(a) Name of JFMC: ____________________ ____________________ ____________________

Forest division: ______________ Range: ______________

Village: _____________

(b) JFMC profile

Particulars	*BPL*	*SC*	*ST*	*Others*	*Total*
Village population					
Village households					
JFM members					
JFM households					
Office-bearer belongs to	ST	SC	OBC	Woman	General
President					
Vice president					
Joint secretary					
Secretary (official/non-official)					

(c) Plantation(s)

Location, legal status, and ownership	*Year*	*Gross area (ha)*	*Plantation net area (ha)*	*Seedlings planted*	*Plantation model*	*Distance (km from village)*

2. Evaluation details

(a) Survival of planted seedlings: sample plots measured _______ (specify number)

Average	Species			Species			Others		
	Ht<0.3 m	0.3–1.37 m	>1.37 m	Ht<0.3 m	0.3–1.37 m	>1.37m	Ht<0.3 m	0.3–1.37 m	>1.37 m
Number per hectare									
Survival percen-tage									

List out the species:
Status of the plantation journal:

(b) Health of natural regeneration, coppice, and seedlings from sowing: sample plots measured ________ (specify number)

Average	Natural regeneration		Coppice regeneration		Seedlings from sowing seeds		
	0.3–1.37 m	>1.37 m	0.3–1.37 m	>1.37 m	Ht <0.3 m	0.3–1.37 m	>1.37 m
Number per hectare							

(c) Microplanning: observations on community participation and ownership (PRA/RRA and examination of JFMC records)

(i) Preparation
(ii) Language of microplan
(iii) Is it available at the village level: Yes/No
(iv) Implementation including review and deviations

(d) Village development fund (PRA/RRA and examination of JFMC records)

Month and year of commencement	Total amount	Current balance	How many types of activities	Beneficiary households

(e) Inclusive growth of village (PRA/RRA and examination of JFMC records)

(i) FDA activities (including entry-point activities) and employment generated

Activities and duration	*Expenditure*	*Employment generated (MD)*	*Beneficiary households*	*BPL households*	*SC/ST households*

(ii) Observations on self-help groups, forest-based enterprises, NGOs, and insurance in the village

(iii) Benefit flows from forests and FDA, including entry-point activities and capacity building

Benefits generated	*Quantity and unit*	*Households benefited*	*BPL households*	*SC/ST households*

(iv) Observations on communication about NAP, its changes, and transparency of fund use

(f) Fund flow and fund utilization (PRA/RRA and examination of JFMC records)

Particulars	*Amount*	*Date*	*Remarks*
Grant from NAEB to FDA (based on JFMC/FDA records)			
Transferred to JFMC by FDA (based on JFMC/FDA records)			
Funds not utilized		NA	
Bank balance		Date	
Audit agency (attach copy of the latest audit observations)		Date	
Payment delayed during the evaluation period			
		Number of cases	
Beyond 15 days due to non-availability of grant			
Activities delayed due to non-availability of grant			

(g) Coordination with PRI and other agencies (observations based on PRA and RRA)

(h) FDAs contributing to human development and community happiness (from household survey)

Parameters	*Households sample size*	*Households (no response)*	*Average rating (respondents)*
Level of satisfaction			
Household economy saving, debt, and power			
Infrastructural facilities			
Health and hygiene			
Quality of forests and environment			
Participation and conflict management			
Gender equality			
Use of timber from the village area by outsiders			
Interaction with government and other organization			
Benefits from soil and moisture conservation works			
Level of use			
Firewood as primary energy source for cooking			
Agriculture as principal occupation			
Engaged in NTFP, firewood, and timber collection			
Ownership of pressure cooker and pan			
Proportion of all masonry house			
Changes in use of forests by wildlife			

(i) Percentage of households having savings accounts

Bank		*Post office*	

3. Other related critical observations

4. Overall assessment

Place: (Signature of evaluator)

Date: Name:

Organization:

ACKNOWLEDGEMENTS

The researchers are thankful for the support and funding provided by the NAEB and the MoEF, Government of India. We welcome the valuable suggestions and comments given by the review committee in improving this document. We are also thankful to the forest officers for their valuable discussions and feedback in finalizing the evaluation format. We are thankful to all the concerned forest officials and personnel in the Nagpur Circle, Maharashtra Forest Department, Bhopal, Betul, and Mandla Circles of Madhya Pradesh Forest Department for their valuable support in carrying out the fieldwork in testing the various sampling procedures. Last but not the least, we thank all the concerned persons for their support at the Indian Institute of Forest Management, Bhopal.

REFERENCES

NAEB (National Afforestation and Eco-development Board). 2005. *Compendium of Operational Guidelines and Circulars* (Revised as on December 2005). New Delhi: NAEB, Ministry of Environment and Forests, Government of India

Robinson, A. P., D. C. Hamlin, and S. E. Fairweather. 1999. Improving the forest inventories. *Journal of Forestry* 1999 (DEC): 38–42

ANNEX 6.1

The following is the evaluation format for the first concurrent evaluation of forest development agency projects sanctioned under the National Afforestation Programme (NAP).

EXECUTIVE SUMMARY

- Brief summary of the report, cross-linking relevant portions in the main text
- Evaluation findings summarized as overall grading of the project (Table 6.1A)

PROJECT EVALUATION PRO FORMA FOR NAP

The information and data required for project evaluation have a strong bearing on the "Approved Project Proposal" and "Periodic Progress Reports" submitted by the project implementing FDAs to the NAEB. Therefore, the first task of the evaluating agencies would be to secure relevant information from the FDAs and conduct the evaluation in the light of the following criteria.

TABLE 6.1A Grading a project on a scale of 1 to 10

Quantitative aspect	Physical
	Financial
Qualitative aspect	Survival
	Health of plantations
	Maintenance
	Sustainability
JFM	Degree of people's participation
	Satisfaction over perceived benefits by the local population
FDA mechanism	Composition of general and executive bodies of the FDA
	Role of FDA in administrative, supervisory, and monitoring mechanism of the project
	Flow of funds from FDA to VFCs
	Planning, implementation, and maintenance of assets directly by VFCs/EDCs
Overall grading of the project	Outstanding (8–10)

Note The evaluator will provide a clear, consistent, concise, and coherent justification for assigning scores (1–10) for the various parameters listed for grading the FDA project.

A. GENERAL (PROJECT PROFILE AND PROPOSED ACTIVITIES)

The evaluating agencies should provide background information and salient features of the project. This should include the following information.

A.1 Project Title

- Project title
- Project implementing FDA: district, forest division, forest circle, state
- Project location(s): sub-watersheds, development blocks, forest ranges, associated villages
- Project cost, duration, and area

A.2 Project Proposal

- Introduction
- Project areas and locations
- Any previous project
- Ongoing project: socio-economic profiles of villages covered, land holding and distribution, land use pattern, watershed features, forests covered, degree of pressure on forest resources
- Project objectives
- Project components
- Salient features, including proposed strategies for treatment
- Creation of JFMC and development funds
- Implementing agencies: composition and constitution of FDA and field implementing units
- Microplanning and JFM
- Entry-point activities
- Approved programme of works

B. QUANTITATIVE ANALYSIS

B.1 Classification of Area Covered/treated

- Provide separate table for each financial year (FY) from year of sanction
- Total coverage (Table 6.1B) so far

B.2 Physical and Financial Achievements

The data should be compiled in subheads in the following format.

TABLE 6.1B Total area covered/treated

Category	*Added natural regeneration*	*Artificial regeneration*	*Slivopasture*	*Bamboo plantation*	*Cane plantation*	*Mixed plantation*	*Regeneration of perennial herbs/shurbs*	*Total*	*Per cent of total project area*
Forest land									
Community land									
Revenue land									
Government wasteland									
Private land									
Others									
Total									

- The coverage (Table 6.1C) so far from the date of sanction
- Table for each FY from year of sanction
- Component-wise summary of physical area (sites) treated in different forest/ ranges/sub-watersheds in each year.
- Adequacy of the planting stock raised or available in each year: summary either forest range or nursery-wise.

TABLE 6.1C Total coverage since the sanction date

Item	*Physical (area in ha)*		*Financial (rupees in lakh)*	
	Target for the year	*Achieved in the year*	*Target for the year*	*Achieved in the year*
Natural regeneration				
(a) Advance work				
(b) Creation				
(c) Maintenance				
Subtotal				
Management intervention				
(a) Advance work				
(b) Creation				
(c) Maintenance				
Subtotal				
Pasture/silvopasture development				
(a) Advance work				
(b) Creation				
(c) Maintenance				
Subtotal				
Soil and moisture conservation				
Extension/awareness/ JFM/training				
Fencing				
Monitoring and evaluation				
Improved technologies				
Overheads				
Total				

B.3 People's Participation

The data should be compiled in the following subheads.

B.3.1 Participatory rural appraisal/microplanning

- Inspection of records and comments whether process of microplanning in the project is "to be initiated", "ongoing", or "completed".
- Expenditure incurred on the preparation of microplans (rupees in lakh), as also percentage of the total approved project outlay.

B.3.2 Entry-point activities

The assets created under entry-point activities must be recorded. Provide a list of entry-point activities village wise. The expenditure incurred in each FY on entry-point activities should be presented as in Table 6.1D.

TABLE 6.1D Details of entry-point activities

Name of village	*Assets created under FDA*	*Number of assets*	*Expenditure incurred*	*People's contribution towards expenditure*

Entry-point assets in the project area should be created. Brief comments on their quality, effectiveness, and usefulness to the project, along with their impact on the local people, including motivating them to willingly participate in the protection of forest areas, should be noted.

B.3.3 Joint forest management

For comments on the JFM approach adopted in the project, provide the following information.

- Number of villages where formation initiated in the division
- Number of villages where formal committees established in the division
- Total number of committees established so far in the project area
- Total membership in terms of households/families
- Percentage of women, SCs, and STs in the JFMCs (Table 6.1E)

TABLE 6.1E Representation of SCs, STs, and others in JFMCs

	Male	*Female*	*Total*
Scheduled caste			
Scheduled tribe			
Others			
Total			

- Establishment of village development fund: Yes/No
  If yes, quantum of fund available as on __-__-20__
- Joint forest management committee meetings/forest development agency meetings (Table 6.1F).

TABLE 6.1F Details of JFMC and FDA meetings

	JFMC meetings	*FDA meetings*
Number of general body meetings held		
Number of executive body meetings held		

- Microplanning
  Prepared/discussed/ratified by VFC/under implementation
- List of important registers/records maintained by and available with
  - JFMCs
  - FDAs

B.3.4 Capacity building

Training of JFMC/community participation conducted (Table 6.1G). Provide separate table for each FY from year of sanction.

TABLE 6.1G Details of capacity building activities

Training programme conducted			
Staff			
Number trained			
Topics covered			
Community			
Number trained			
Topics covered			
Organizations involved (trainers)			
Year-wise expenditure (rupees in lakh) incurred on training activities			
Year 1	Year 2	Year 3	**Total**

B.3.5 Monitoring and evaluation

- Number of inspection by
  - Chief conservator of forests
  - Conservator of forests

- Deputy conservator of forests
- Whether FDA conducted regular monitoring and evaluation of project activities (give the expenditure incurred as in Table 6.1H).

TABLE 6.1H Expenditure incurred on monitoring and evaluation by year

Year 1	*Year 2*	*Year 3*	*Total*

- Were any independent agencies involved in the above monitoring process? If yes,
  - Provide the list of agencies, copies of reports, and list of main findings/recommendations
  - Frequency of monitoring and/or evaluation
  - Number of such activities carried out in each FY (as in Table 6.1I)

TABLE 6.1I Number of monitoring and evaluation activities in each FY

Year 1	*Year 2*	*Year 3*	*Total*

B.4 FDA Mechanism

- Outline the constitution of FDA, that is, composition of general and executive bodies, along with deviations, if any, reasons thereof, and concurrence of MoEF thereto.
- Highlight the strengths and weaknesses, if any, pertaining to mutual responsibilities assigned to FDA (administrative, supervisory, and monitoring) and VFCs/EDCs (execution of fieldworks).
- Comment on whether the project implementing VFCs/EDCs are also being benefited through any other forestry/community development project from state/central government. If not, then what role and initiatives the FDA proposes in regard to the overall rural/eco-development of the project areas?
- Provide a brief outline of the FDA bank account, its mode of operation, auditing status supported with progress of works as envisaged in the guidelines. Comment on the flow of funds to the VFCs/EDCs and disbursement of funds by JFMCs for works executed. If delayed, give reasons thereof.

- Similar details should also be provided for the individual VFC/EDC whose works were assessed and interactions were held. Provide details of statements of account and progress reports sent by VFC/EDC to the FDA. Comment on the receipt of funds by VFCs/EDCs.

C. QUALITATIVE ANALYSIS

The evaluator will compile information on the socio-economic profile and forest/afforestation details of all NAP–JFMCs in the FDA and randomly select a statistically significant number of JFMCs for in-depth study based on a stratified random sample. The whole design should be statistically sound so that there is least human bias. The sampling design should be described in detail, in addition to the information in the prescribed format for the evaluation report.

C.1 Assessment of Plantations, SMC Works, and their Output

- Brief comment on the suitability of areas selected for project implementation.
- Assessment of work as seen in the field, including quality of work.
- Survival rate (estimated statistically, indicating the methodology adopted and average height of plants).
- Assessment of regeneration/rehabilitation status of degraded lands treated.
- Technology or methodology adopted under each of the components sanctioned by the NAEB; an analysis of the technology used and its effectiveness in brief.
- Increase in the availability of biomass (fuelwood, fodder, other NTFPs, and small timber).
- Improvement in the quality of life of people in project area (status of health and education, availability of drinking water, increase in per capita income).

C.2 People's Participation

- Has the choice of species been decided in consultation with the local people? Also comment briefly whether these are in consonance with the local needs.
- Briefly comment on the suitability of species taken up for NTFP and medicinal plants (perennials) from the economic point of view.

- Has the project created awareness among the local population about the responsibilities they expected to discharge and the benefits that are likely to accrue from the project?
- Approach adopted by the project authorities for motivating the public.
- Have the means of communication adopted been successful and has there been a favourable response?
- Have the village communities been involved in
  - Project planning.
  - Project implementation.
  - Usufructs sharing mechanism.
- Are people aware of the government order issued (if any) with regard to JFM/usufructs sharing mechanism?
- Has any formal arrangement been evolved for sharing the intermediate and final produce of the project activities?
- Evaluation and comments on the effectiveness of the mechanism evolved and the equitableness of the same.
- What role has been played by the local communities/committee in arriving at the mechanism?

C.3 Assets: Protection and Maintenance

What measures have been used for protecting the assets created in the project? State the quality and quantity of measures adopted and their effectiveness. The evaluator may state whether in spite of the measures adopted, the plantations are subjected to grazing or other abiotic damages. State whether these issues were placed before the VFC? If so, what were the results? If not, why? Whether the VFC has been effective in sorting out these issues?

C.4 Maintenance of Records

- Have the records of nurseries indicating the number and species of seedlings raised been maintained?
- Are they adequate and posted from time to time? If not, state the reasons and effects of unposted records on the implementation of project.
- Have plantation journals been maintained and posted up to date? Has a surveyed sketch of the plantation been pasted on the register?
- Have the inspecting officials of the implementing agency, such as divisional forest officer (DFO), conservator of forest (CF), or chief conservator of forest (CCF), recorded their observations/comments?

- Have the records of VFC/VDA General Body and Executive Body meetings been maintained?

C.5 Project Benefits

Comment on whether implementing authorities have quantified the tangible and intangible benefits that the regenerated areas would be able to offer to the local people. Or else, state whether a suitable and appropriate system to forecast, evaluate, and monitor such benefits is being developed. If not, then give reasons thereof.

Provide a brief outline of the employment generated (man days) by the project activities, especially afforestation, soil and moisture conservation works, and entry-point activities in each VFC/EDC. Also comment on the extent of such benefits having gone directly to the local women/SCs/STs.

C.6 Project Reporting Activities

- What returns (reports) have been submitted by the implementing agency to the state/central government? State whether these reports have been sent regularly on due dates. If not, give reasons thereof.
- State whether any state-level agency has monitored project activities. If so, highlight the outcome of such monitoring/evaluations and the comments, if any.
- If no agency evaluated the project activities, then give reasons thereof.

C.7 Project Constraints/Limitations

- What constraints/limitations were faced by the project authorities?
- State whether the fund flow was smooth and without delays.
- If the flow was not smooth, where did the delay arise? This must be checked with records.
- State whether the project authorities made adequate efforts to get the funds released in time.

C.8 Suggestions for Improvements

C.8.1 From implementing/supervising agencies

- State whether there is any scope for improving the project output.
- Have the project authorities felt any need for improving upon any particular activity or methodology?

- Do the people of the area feel any need to improve any particular aspect of the project?

C.8.2 From the evaluating agency/consultants

- State whether the project should be continued/modified/discontinued (give reasons).
- Give any other relevant recommendation(s).
- The evaluator will discuss the evaluation design and draft outcomes with functionaries of FDA (DFO, CF, and CCF) before the start, and upon conclusion of the study. Provide the names and designations of the functionaries of FDA (DFO, CF, and CCF) with whom the evaluator has interacted, along with the date and appraisal summary of discussions with the officers.

ANNEX 6.2

TABLE 6.2A Household questionnaire at JFMC Level

Forest division: *Household number:*		*Range:* *Name:*			*Village:*	
Type of house living						
Pucca	Decent kutcha	Thatched	Dilapidated	Makeshift		
Level of satisfaction						
A. Household economy—savings, debt, and power						
1. Savings and debt	Enough to meet future needs	Some savings for immediate needs	No savings and no debt	Occasional debt	Annual cycle of debt	Perpetual debt
2. Voice and power	Able to influence village opinion	Openly express our views	Express within closer groups	Express through others/groups	Our opinion has no value	Fearful to express our opinion
3. Support you get from neighbours in emergency				Always	Sometimes	Never
B. Infrastructure facilities						
1. Infrastructure facilities used by household				Good	Average	Poor
2. Infrastructure facilities at the village level				Good	Average	Poor
C. Health and hygiene						
1. Sanitation facilities				Good	Average	Poor
2. Household facilities				Good	Average	Poor

Contd...

TABLE 6.2A *Contd...*

D. Quality of forests and environment			
1. The condition of the forest and quality	Good	Average	Poor
2. Access and quality of drinking water	Good	Average	Poor
3. Access and quality of irrigation water	Good	Average	Poor
E. Participation and conflict management			
1. How many village-level meetings you or your household members attended	Regularly	Sometimes	Never
2. Resolving the conflicts at village level	Regularly	Sometimes	Never
3. According to you, what is the participation level of villagers related to village issues	Good	Average	Poor
F. Gender equality			
1. How many village-level meetings are attended by women of your house	Regularly	Sometimes	Never
2. What importance is given to women's opinion and their concerns at community-level decisions	High	Medium	Low
G. Use of timber from the village area by outsiders			
1. Use of timber from the village area by outsiders	High	Medium	Low
2. Damage caused by outsiders' use	High	Medium	Low
H. Interaction with government and other organizations			
1. Level of satisfaction with forest department	High	Medium	Low
2. Level of satisfaction with other government departments	High	Medium	Low
3. Level of satisfaction with other non-government departments	High	Medium	Low
4. Benefits from soil and moisture conservation works	High	Medium	Low

Contd...

TABLE 6.2A *Contd...*

I. Level of use			
1. Firewood as primary energy source for cooking	High	Medium	Low or nil
2. Agriculture as principal occupation	Yes	No	
3. Engaged in NTFP collection	Regularly	Sometimes	Never
4. Engaged in firewood collection	Regularly	Sometimes	Never
5. Engaged in timber collection	Regularly	Sometimes	Never
6. Ownership of pressure cooker and pan	Yes	No	
J. Changes in use of forests by wildlife			
1. Use of forest by wildlife has increased since JFM	High	Medium	Low
2. How are you satisfied with the overall functioning of JFMC in your village?	Highly satisfied	Somewhat satisfied	Not at all
K. Savings bank account			
1. Does your household have a savings bank account or post office account: Yes/No			
2. How far is your bank/post office: _____ km			

L. Benefits from JFM programme to your household		
Activity	Number of days/quantity	Total monetary benefit
1. Direct benefit		
2. Indirect benefit		

Has anyone attended any training/awareness programme or exposure visit under JFM? List the events and the number of people participated. Also list the items that are not covered in the table.

ANNEX 6.3

SCORING FORMAT

The criterion and indicator approach of sustainable forest management highlights three critical considerations: (1) ecological, (2) economic, and (3) social. Correspondingly, the FDA activities may also be evaluated on similar lines. The evaluator should assign a separate set of scores to the FDA report and the fieldwork report based on field measurement and observations made during the evaluators' visits to the sample sites. Subjective scoring often has large error margins, and hence use of a smaller scale than the current 10-point scale is recommended. Equal weight is given to ecological, economic, and social considerations in the following scoring format.

SCORING FORMAT FOR EVALUATING NAP AND FDA ACTIVITIES

Each component should be graded by assigning a value on a five-point scale. (Assign zero to indicate absence. Do not use more than one digit after decimal.)

TABLE 6.3A Format of scoring for the evaluation of NAP and FDA activities

Consideration	*Parameters*	*Evaluation scores*		*Component average*
		FDA report	*Fieldwork*	
Ecological	Achievement of physical target		NA	
	Health of plantations, species selection, and suitability	NA		
	Maintenance of natural regeneration and coppice			
	Survival of planted seedlings			
Economic	Achievement of financial target		NA	
	People's satisfaction in the perceived benefits	NA		
	Flow of funds from FDA to the village committee and its use			

Contd...

TABLE 6.3A *Contd...*

Consideration	*Parameters*	*Evaluation scores*		*Component average*
		FDA report	*Fieldwork*	
	Employment generation, VDF, and other supports from FDA			
Social	People's participation in programme implementation and microplanning			
	Representation and concerns for deprived sections and women			
	Level of community satisfaction, quality of life, and SHGs			
Overall sustainability				

ANNEX 6.4

GUIDELINES FOR SELECTING SAMPLE SITES

- List all the sites in the given division to be evaluated along with range, compartment number (if compartment number is not available, use revenue survey number), planting year, and net plantation area (as shown in Table 6.4A).
- Calculate cumulative area (column number 7).
- Divide the cumulative area (column number 7) by the total area of plantation in the division, which is a cumulative per cent area (column number 8).
- Select five sites randomly using the "Table of Random Numbers."

Example: Let us consider sampling of sites for evaluation for FDA plantations in Betul division, Madhya Pradesh (Table 6.4A).

Step 1: List all the sites to be evaluated in serial order according to plantation year.

Step 2: Calculate cumulative areas. The cumulative area of the first site is the same as the plantation area of that site, that is, 50. The cumulative area of the second site is the cumulative area of the first site plus the plantation area of the second site. The cumulative area of the third site is the cumulative area of the second site plus the plantation area of the third site. Continue the procedure for all other entries

Step 3: Calculate the cumulative per cent area of each site by dividing the cumulative area of each site with the total planted area.

Cumulative per cent area = (Cumulative planted area × 100)/ Total planted area

For site 1, cumulative per cent area is $(50 \times 100)/1700 = 2.9\% \approx 3\%$; for site 2, cumulative per cent area is $(100 \times 100)/1700 = 5.8\% \approx 6\%$. Continue the procedure for all the remaining entries.

Step 4: Select one of the random number tables, and select any column or row. We selected the "Table of Random Numbers 2" and used the 11th column. The first five random numbers are 94, 99, 05, 43, and 67 for selecting five sites. For each of these numbers, we find the associated compartment/site with a cumulative per cent

TABLE 6.4A Selection of sites from Betul division, Madhya Pradesh

S. No.	*Division*	*Range*	*Compartment number*	*Year*	*Area*	*Cumulative area*	*Cumulative per cent area*	*Sites selected for evaluation*
1	Betul	Betul	245	2003/04	50	50	03	
2	Betul	Betul	263	2003/04	50	100	06	Selected
3	Betul	Shapur	192	2003/04	50	150	09	
4	Betul	Bhaira	155A	2003/04	50	200	12	
5	Betul	Bhaira	128	2003/04	50	250	15	
6	Betul	Bhaira	164	2003/04	50	300	18	
7	Betul	Saarni	338	2003/04	50	350	21	
8	Betul	Ranipur	510	2003/04	50	400	24	
9	Betul	Ranipur	507	2003/04	50	450	26	
10	Betul	Ranipur	281	2003/04	50	500	29	
11	Betul	Betul	261	2003/04	50	550	32	
12	Betul	Saarni	482	2003/04	50	600	35	
13	Betul	Saarni	481	2003/04	50	650	38	
14	Betul	Ranipur	879	2003/04	50	700	41	
15	Betul	Betul	245A	2004/05	50	750	44	Selected
16	Betul	Betul	257A	2004/05	50	800	47	
17	Betul	Shapur	183	2004/05	50	850	50	
18	Betul	Shapur	239	2004/05	50	900	53	
19	Betul	Shapur	238	2004/05	50	950	56	
20	Betul	Ranipur	280	2004/05	50	1000	59	

Contd...

TABLE 6.4A *Contd...*

S. No.	*Division*	*Range*	*Compartment number*	*Year*	*Area*	*Cumulative area*	*Cumulative per cent area*	*Sites selected for evaluation*
21	Betul	Betul	272	2004/05	50	1050	62	
22	Betul	Bhaira	130A	2004/05	50	1100	65	
23	Betul	Saarni	P484	2004/05	50	1150	68	Selected
24	Betul	Shapur	243	2005/06	50	1200	71	
25	Betul	Shapur	223	2005/06	50	1250	74	
26	Betul	Bhaira	139A	2005/06	50	1300	76	
27	Betul	Saarni	334	2005/06	50	1350	79	
28	Betul	Saarni	340	2005/06	50	1400	82	
29	Betul	Bhaira	385	2005/06	50	1450	85	
30	Betul	Shapur	P226	2006/07	50	1500	88	
31	Betul	Bhaira	95	2006/07	50	1550	91	
32	Betul	Saarni	357	2006/07	50	1600	94	Selected
33	Betul	Ranipur	R503	2006/07	50	1650	97	
34	Betul	Saarni	R346	2006/07	50	1700	100	Selected

Note Total planted area is 1700.

area larger than or equal to each of the number. In Table 6.4A, the random number 43 will be used to select plantation at serial number 15—compartment number 245A—as sample because it is greater than the cumulative per cent area (41%) for plantation at serial number 14, but less than the cumulative per cent area (44%) for this plantation.

Step 5: If a particular site is already selected as sample using a random number, and if any other random number also indicates selecting the same, add the additional random number to the random table sequence shown in step 4.

Step 6: Repeat steps 4 and 5 till the required number of sites is selected.

ANNEX 6.5

GUIDELINES FOR STRIP SAMPLING FOR MEASURING SURVIVAL PERCENTAGE

Step 1: To carry out strip sampling, stratify the selected sample site by topography and soil type. If these are large in heterogeneity, the entire area should be subdivided into smaller sections. Survival percentage should be carried out for each section and summed together.

Step 2: Strip width should be three times the width of spacing between the plants.

Step 3: Documentation should be carried out in segments double the segments width.

Step 4: Counting should be carried out for every third segment along the strip. Incomplete segment should not be counted.

Step 5: The first strip may begin at any random point. Subsequently, roughly parallel strips may be repeated in such a manner, on both sides of the first strip, such that the entire area is covered. If roughly parallel strips are spaced out at six times the strip width, the sampling exercise could be carried out faster with sampling intensity in the range of 4%–6%.

Step 6: For the purpose of reporting, the survival per cent should be weighted by net area planted in the same model. The percentage should be reported separately for each plantation model covered for the counting and compared with the value reported by the FDA. The survival count of individual plantations should be calculated by weighted net plantation area and average number of seedlings in each model.

Note The details of mapping of the selected site through Global Positioning System (GPS) should be carried out according to the manual of the GPS to be used. (The instructions would vary from one model to another.)

ANNEX 6.6

GUIDELINES FOR HOUSEHOLD SAMPLING SURVEY

- Carry out the household survey in the village under which the selected sampled plantation site falls under ransom sampling.
- Use the latest updated member household list to select households. (Give numbering to all the households from serial number one to the last number.)
- Use one of the random number tables (Annex 6.10) for selecting the households with the help of the following guidelines.
  - If the village has up to 20 households, carry out household interviews in all the households. In case total households are around 100, select 20 households using the random number table.
  - Divide 3000 by the number of households, and round off the value to the nearest integer to obtain the sampling per cent. Select equal number of samples from each complete block of 100 and proportionate integer value for an incomplete 100.
  - If total households are more than 750, use 4% as sampling per cent. Select four samples from each completed block of 100 and proportionate integer value for an incomplete 100. Table 6.6A shows the sample size needed for villages of different sizes.

  For example, a selected village may have 245 households. The sampling intensity will be $3000/245 = 12.24 \approx 12\%$. Select 12 samples from 1 to 100. Select other 12 from 101 to 200. Select other 5 ($45 \times 12\%$) samples from 201 to 245. Thus, the total households to be interviewed will be 27.

  In another example, consider the total households in a selected village to be 537; the sampling intensity will be $3000/537 = 5.58 \approx 6\%$. Select six samples from each of the completed blocks of 100 (1–100, 101–200, 201–300, 301–400, and 401–500), and add two ($35 \times 6\%$) samples from 501 to 535. Thus, the total households to be interviewed will be 32.

  In yet another example of a village having total of 870 households, select four samples from each of the completed blocks of 100 and add three ($70 \times 4\%$) from 801 to 870. Thus, the total households to be interviewed will be 35.

TABLE 6.6A Sample size of households (at village level)

S. No.	Total households in a village	Sample size	Number of houses for each 100
1.	5 to 20	All houses	
2.	21 to 100	20	
	101 to 750	Around 30	Proportionately from each block of 100
3.	More than 750	4%	Four from each block of 100

Example 1: Let us consider that a village has 100 houses and are numbered from 1 to 100 (Table 6.6B). Select Random Number Table 1 or Random Number Table 2 and use any row or column of the selected table. Let us select the third column of Random Number Table 2. The numbers are 15, 32, 32, 72, 38, 61, 25, 56, 49, 18, 70, 18, 10, 82, 75, 49, 47, 51, 77, 31, 25, 63, 07, 40, 96, 78, 83, 49, 81, 71, 30, 29, 19, 39, 68, 50, 38, 95, 97, and 51.

In the above numbers, if any number occurs more than once, select that number only once. For example, the number 32 occurred as the second and third number. Select 32 as the second random number and 72 as the third random number. Continue this process until you get 20 random numbers (from Random Number Table 2). Thus, the first 20 random numbers (as highlighted in Table 6.6B) selected are as follows: 15, 32, 72, 38, 61, 25, 56, 18, 70, 10, 82, 75, 49, 47, 51, 77, 31, 63, 07, and 40.

TABLE 6.6B Village houses numbered from 1 to 100

01	11	21	**31**	41	**51**	**61**	71	81	91
02	12	22	**32**	42	52	62	**72**	**82**	92
03	13	23	33	43	53	**63**	73	83	93
04	14	24	34	44	54	64	74	84	94
05	**15**	**25**	35	45	55	65	**75**	85	95
06	16	26	36	46	**56**	66	76	86	96
07	17	27	37	**47**	57	67	**77**	87	97
08	**18**	28	**38**	48	58	68	78	88	98
09	19	29	39	**49**	59	69	79	89	99
10	20	30	**40**	50	60	**70**	80	90	100

The households appearing at the corresponding serial numbers in the list are the sample households. The evaluator will visit the respective households to conduct interviews and collect information according to the format given in the Village Household Schedule. If an adult person of the selected household is not available for the household interview, the evaluator should carry out the interview in the adjoining household (replace that sample household with the adjoining household). If the household on the right side of the sample household is selected as the first replacement, all other replacements must be done in a similar fashion (that is, on the right side of the sample household).

Example 2: Let us consider that 250 households in a village are numbered from 1 to 250 (Table 6.6C). In this case, the total household size of the village is 250 and the sampling per cent will be 3000/240 = 12%. We have to select 12 random numbers from 1 to 100, another 12 numbers from 101–200, and 6 numbers from 201 to 250. Select any row or column from Random Number Table 1 or Random Number Table 2. Let us select the sixth row in Random Number Table 1. The numbers are 28, 11, 45, 19, 87, 39, 13, 73, 17, 33, 39, 38, 63, 26, 70, 33, 10, and 28. Since the total numbers are 18 in this row, which is less than the required number 30, continue selecting the numbers from the immediate next row by moving down or from the immediate previous row, by moving up, from right to left. In this case, we select the numbers from the immediate next row by moving down. The numbers of the seventh row are 78, 81, 30, 59, 37, 02, 25, 32, 70, 66, 08, 65, 46, 26, 88, 20, 17, and 49. The random numbers from the sixth and seventh rows after eliminating repeated numbers are as follows: 28, 11, 45, 19, 87, 39, 13, 73, 17, 33, 38, 63, 26, 70, 10, 78, 81, 30, 59, 37, 02, 25, 32, 66, 08, 65, 46, 26, 88, 20, and 49.

Now the first 12 numbers will be the random numbers to select the serial numbered households from 1 to 100, and the next 12 numbers will be obtained by adding 100 to the next 12 random numbers chosen to select the serial numbered households from 101 to 200. The next six numbers will be obtained by adding 200.

Thus the first 12 sample numbers are 28, 11, 45, 19, 87, 39, 13, 73, 17, 33, 38, and 63. The second 12 numbers are 126, 170, 110, 178, 181, 130, 159, 137, 102, 125, 132, and 166. The last six numbers are 208, 246, 226, 220, and 249 (as highlighted in Table 6.6C).

TABLE 6.6C Village households numbered from 1 to 25

01	21	41	61	81	101	121	141	161	181	201	221	241
02	22	42	62	82	102	122	142	162	182	202	222	242
03	23	43	63	83	103	123	143	163	183	203	223	243
04	24	44	64	84	104	124	144	164	184	204	224	244
05	25	45	65	85	105	125	145	165	185	205	225	245
06	26	46	66	86	106	126	146	166	186	206	**226**	**246**

Contd...

TABLE 6.6C *Contd...*

07	27	47	67	87	107	127	147	167	187	207	227	247
08	28	48	68	88	108	128	148	168	188	**208**	228	248
09	29	49	69	89	109	129	149	169	189	209	229	**249**
10	30	50	70	90	110	130	150	170	190	210	230	250
11	31	51	71	91	111	131	151	171	191	211	231	
12	32	52	72	92	112	132	152	172	192	212	232	
13	33	53	73	93	113	133	153	173	193	213	233	
14	34	54	74	94	114	134	154	174	194	214	234	
15	35	55	75	95	115	135	155	175	195	215	235	
16	36	56	76	96	116	136	156	176	196	216	236	
17	37	57	77	97	117	137	157	177	197	217	237	
18	38	58	78	98	118	138	158	178	198	218	238	
19	39	59	79	99	119	139	159	179	199	219	239	
20	40	60	80	100	120	140	160	180	200	**220**	240	

ANNEX 6.7

GUIDELINES TO CONSOLIDATE THE HOUSEHOLD SURVEY QUESTIONNAIRE

- All the questions except I2, I6, K1, and K2 should be reported in weighted average.
- The questions I2, I6, K1, and K2 should be reported as percentage of households interviewed.
- Coding of responses for questions other than questions numbered I2, I6, K1, and K2 is given in Table 6.7A.

TABLE 6.7A Coding of responses to the household survey questionnaire

Q. No.	*Question*	*Response*	*Code*	*Reported as*
A1	Saving and debt	Enough to meet future needs or some saving for immediate needs	3	Weighted average
		No savings and no debt	2	
		Occasional debt, annual cycle of debt, or perpetual debt	1	
A2	Voice and power	Able to influence village opinion, openly express our views, or express within closer groups	3	Weighted average
		Express through others/goups	2	
		Our opinion has no value or fearful to express opinion	1	
A3	The support you get from neighbours in emergency	Always	3	Weighted average
		Sometimes	2	
		Never	1	

Contd...

TABLE 6.7A *Contd...*

Q. No.	Question	Response	Code	Reported as
B1	Infrastructure facilities used by household	Good	3	Weighted average
		Average	2	
		Poor	1	
B2	Infrastructure facilities at the village level	Good	3	Weighted average
		Average	2	
		Poor	1	
C1	Sanitation facilities	Good	3	Weighted average
		Average	2	
		Poor	1	
C2	Health of household	Good	3	Weighted average
		Average	2	
		Poor	1	
D1	The condition of forest and quality	Good	3	Weighted average
		Average	2	
		Poor	1	
D2	Access and quality of drinking water	Good	3	Weighted average
		Average	2	
		Good	1	
D3	Access and quality of irrigation water	Good	3	Weighted average
		Average	2	
		Poor	1	

Contd...

TABLE 6.7A *Contd...*

Q. No.	Question	Response	Code	Reported as
E1	How many village-level meetings you or your households members attended	Regularly	3	Weighted average
		Sometimes	2	
		Never	1	
E2	Resolving the conflicts at village level	Regularly	3	Weighted average
		Sometimes	2	
		Never	1	
E3	In your view, what is the participation level by villagers for village issues	Good	3	Weighted average
		Average	2	
		Poor	1	
F1	How many village-level meetings are attended by women of your house	Regularly	3	Weighted average
		Sometimes	2	
		Never	1	
F2	What importance is given to women's opinion/concerns in community decisions	High	3	Weighted average
		Medium	2	
		Low	1	
G1	Use of timber from the village area by outsiders	High	3	Weighted average
		Medium	2	
		Low	1	
G2	Damage caused by outsiders' uses	High	3	Weighted average
		Medium	2	
		Low	1	

Contd...

TABLE 6.7A *Contd...*

Q. No.	*Question*	*Response*	*Code*	*Reported as*
H1	Level of satisfaction with forest department	High	3	Weighted average
		Medium	2	
		Low	1	
H2	Level of satisfaction with other government departments	High	3	Weighted average
		Medium	2	
		Low	1	
H3	Level of satisfaction with other non-governmental organizations	High	3	Weighted average
		Medium	2	
		Low	1	
H4	Benefits from soil and moisture conservation works	High	3	Weighted average
		Medium	2	
		Low	1	
I1	Firewood as primary energy source for cooking	High	3	Weighted average
		Medium	2	
		Low or nil	1	
I2	Agriculture as principal occupation		%	Percentage
I3	Engaged in NTFP collection	Regularly	3	Weighted average
		Sometimes	2	
		Never	1	
I4	Engaged in firewood collection	Regularly	3	Weighted average
		Sometimes	2	
		Never	1	

Contd...

TABLE 6.7A *Contd...*

Q. No.	Question	Response	Code	Reported as
I5	Engaged in timber collection	Regularly	3	Weighted average
		Sometimes	2	
		Never	1	
I6	Ownership of pressure cooker and pan		%	Percentage
J1	Use of forest by wildlife has increased since JFM	High	3	Weighted average
		Medium	2	
		Low	1	
J2	How are you satisfied with the overall function of JFMC in your village	Highly satisfied	3	Weighted average
		Somewhat satisfied	2	
		Not at all	1	
K1	Direct benefits (for each activity reported)	Number of days and quantity	%	Percentage
		Monetary benefit	%	Percentage
K2	Indirect benefit (for each activity reported)	Number of days and quantity	%	Percentage
		Monetary benefit	%	Percentage
K3	Proportion of all masonry house	Pucca, decent kutcha	3	Weighted average
		Thatched	2	
		Dilapidated, makeshift	1	

CALCULATION OF WEIGHTED AVERAGE

Let 25 households be interviewed in a given village. The response to question A3 (the support you get from neighbours in emergency) for the entire village is given in Table 6.7B.

TABLE 6.7B Details of support from neighbours in emergency

Response category	*Good*	*Average*	*Poor*
Code	3	2	1
Number of households	7	9	9

The weights in weighted average are the number of households given Good, Average, and Poor.

Weighted average for support you get from neighbours in emergency

$$= \frac{3 \times 7 + 2 \times 9 + 1 \times 9}{7 + 9 + 9} = 48/25 = 1.92$$

Calculate the weighted average for all the remaining questions in the same manner (refer to Table 6.7A).

In case of percentages, no weighted average is applicable. Report the actual percentage.

ANNEX 6.8

TABLE 6.8A Guidelines for scoring FDA

Parameters	*Score 1*	*Score 2*	*Score 3*	*Score 4*	*Score 5*
1. Ecological parameters					
1.1 Physical achievement of plantation area (per cent of target area covered)	<25%	25%–49%	50%–66%	67%–74%	>74%
1.2 Health of plantations, suitability, and selection of species	Planted species are technically not suitable and are not desired by the local people	Poor condition, but planted species are technically suitable and locally desired	Slow growth, but planted species are technically suitable and locally desired	Moderate growth, and planted species are technically suitable and locally desired	Excellent growth, and planted species are technically suitable and locally desired
1.3 Maintenance of coppice and natural regeneration (per cent of target area)	<25%	25%–49%	50%–66%	67%–74%	>74%
1.4 Weighted survival per cent of planted seedlings (measured survival multiplied with agro-climate specific factor as suggested by the states and sanctioned by the NAEB; default 1.0)	<25%	25%–49%	50%–66%	67%–74%	>74%

Contd...

TABLE 6.8A *Contd...*

Parameters	*Score 1*	*Score 2*	*Score 3*	*Score 4*	*Score 5*
2. Economic parameters					
2.1 Financial achievement (expenditure as percentage of sanctioned budget)	<25%	25%–49%	50%–66%	67%–74%	>74%
2.2 People's satisfaction for the perceived benefits (current and anticipated)	Discontentment and conflict with local forest officials		Weighted average less than 1.5, but regular flow of benefits	Weighted average more than 1.5, and regular flow of benefits	Weighted average more than 2.5, and regular flow of benefits
2.3 Flow of funds from FDA to the village committee and its use (funds received, utilized, and more than half of village committees submitting the utilization reports)	Delays in use not more than six months; reported within three months	Delays in use not more than four months; reported within two months	Delays in use not more than three months; reported within a month	Delays not more than two months	Within stipulated time
2.4 VDF, employment generation, and other supports from FDA for non-forest village development activities	VDF absent; no employment; and limited support to non-forest development works	VDF absent; some generation of employment; and support to non-forest development works	VDF absent; some generation of employment; and support to non-forest development works	VDF present; generation of employment significant; and support to non-forest development works	Effective use of VDF, adequate generation of employment, and some enterprise development

Contd...

TABLE 6.8A *Contd...*

Parameters	*Score 1*	*Score 2*	*Score 3*	*Score 4*	*Score 5*
3. Social parameters					
3.1 People's participation in programme implementation and microplanning (village population participating in the microplanning, in percentage)	<25%	25%–49%	50%–66%	67%–74%	>74%
3.2 Representation and concerns for deprived sections and women (use state government guidelines)	Representation lacking and their voice not considered	Represented according to guidelines but concerns rarely addressed	Represented and concerns partially addressed	Represented according to guidelines and concerns are addressed well	Represented suitably and concerns addressed well
3.3 Level of community satisfaction, quality of life, and SHGs	Pooled average of A1 to D3 at less than 1.0	Pooled average of A1 to D3 at 1.0–1.49	Pooled average of A1 to D3 at 1.5–1.99; and SHGs present	Pooled average of A1 to D3 at 2.0–2.5; and active SHGs	Pooled average of A1 to D3 at least 2.5; and active SHGs

ANNEX 6.9

SCORING FORMAT FOR EVALUATING NAP AND FDA ACTIVITIES

Each component should be graded by assigning a value on a five-point scale. (Assign zero to indicate absence.)

TABLE 6.9A Format of scoring for the evaluation of NAP and FDA activities

Considerations	*Parameters*	*Evaluation scores*		*Component*
		FDA report	*Fieldwork*	*Average*
Ecological	Achievement of physical target	5	NA	
	Health of plantations, species selection, and suitability	NA	3	
	Maintenance of natural regeneration and coppice	3	4	
	Survival of planted seedlings	3	4	
	Component average	(5 + 3 + 3)/3 = 3.67	(3 + 4 + 4)/3 = 3.67	(5 + 3 + 3 + 3 + 4 + 4)/6 = 3.67
Economic	Achievement of financial target	5	NA	
	People's satisfaction for the perceived benefits	NA	3	
	Flow of funds from FDA to the village committee and its use	4	3	
	Employment generation, VDF, and other supports from FDA	3	4	
	Component average	(5 + 4 +3)/3 = 4.00	(3 + 3 + 4)/3 = 3.33	(5 + 4 + 3 + 3 + 3 + 4)/6 = 3.67

Contd...

TABLE 6.9A *Contd...*

Considerations	*Parameters*	*Evaluation scores*		*Component*
		FDA report	*Fieldwork*	*Average*
Social	People's participation in programme implementation and microplanning	3	3	
	Representation and concerns for deprived sections and women	4	3	
	Level of community satisfaction, quality of life, and SHGs	4	3	
	Component average	(3 + 4 + 4)/3 = 3.67	(3 + 3 + 3)/3 = 3.00	(3 + 4 + 4 + 3 + 3 + 3)/6 = 3.33
Overall sustainability	Overall average	3.78	3.33	3.56

Note Calculate the average for each of the ecological, economic, and social considerations as reported by FDA and as documented during fieldwork. Divide the sum of these averages by three to obtain the overall sustainability. Do not use more than one digit after decimal for scores and two digits after decimal for score averages.

ANNEX 6.10

Random Number Table 1

27	27	37	71	99	29	85	56	89	21	48	64	66	31	77	56	32	13
82	11	04	79	31	54	68	88	01	32	05	68	62	88	41	51	94	74
17	14	38	09	96	81	77	43	38	60	39	12	98	99	92	64	37	93
16	79	02	83	89	80	28	58	20	53	01	32	30	18	83	25	75	97
92	08	62	21	04	57	89	64	56	35	41	12	21	63	95	85	65	46
28	11	45	19	87	39	13	73	17	33	39	38	63	26	70	33	10	28
78	81	30	59	37	02	25	32	70	66	08	65	46	26	88	20	17	49
07	39	76	57	29	90	59	26	78	54	33	04	79	64	07	91	88	80
95	56	65	16	15	56	75	37	57	29	65	31	80	72	98	46	83	87
36	01	96	03	67	03	80	06	51	39	24	36	29	95	48	29	92	48
68	37	32	63	84	44	50	76	10	29	41	36	48	38	12	03	80	96
75	29	16	94	18	98	80	66	31	67	79	08	18	19	44	77	50	35
39	11	79	55	53	24	86	16	14	08	74	56	35	92	64	04	76	38
04	37	19	65	72	56	57	66	58	80	36	08	24	05	34	88	77	75
09	61	65	67	95	78	86	25	40	91	79	76	65	79	90	11	05	33
33	15	08	16	09	52	74	93	32	28	64	35	45	06	50	47	25	87
91	84	39	34	88	05	71	02	42	76	18	66	09	81	27	06	85	23
35	54	48	37	23	46	28	69	75	69	19	63	56	26	04	52	47	80
35	70	71	16	16	66	06	85	49	73	66	70	32	75	43	58	40	98
84	43	91	51	47	48	50	44	16	84	25	90	88	88	79	18	44	79
23	94	04	04	81	14	21	05	14	83	56	30	72	85	21	98	99	02
07	06	88	19	60	07	61	12	10	37	88	73	29	58	14	82	24	53
12	79	21	34	77	95	34	71	88	82	03	56	94	62	76	36	76	64
14	92	59	83	02	55	19	21	99	98	81	47	37	48	96	61	98	24
20	65	86	22	59	40	61	37	58	65	92	66	35	31	05	47	35	11
88	14	18	19	40	13	57	32	59	43	24	09	21	86	21	17	21	73
73	99	23	25	35	98	08	67	25	99	48	15	32	07	68	63	56	95
24	14	10	99	35	49	79	40	63	11	05	19	62	44	67	44	46	04
20	11	20	02	33	78	85	77	55	63	55	46	59	64	38	26	50	16
86	54	78	38	11	73	63	38	67	57	32	80	02	70	19	29	11	70
89	25	65	48	66	03	15	16	41	84	99	94	20	36	69	83	77	81
96	41	42	17	91	70	53	40	53	01	04	74	10	09	31	13	38	55
68	74	68	59	12	15	51	37	71	54	48	85	22	98	14	99	72	41
68	20	95	59	32	07	63	95	73	44	88	06	28	43	65	57	89	12

Contd...

Random Number Table 1

17	48	25	02	30	63	56	62	73	24	86	50	65	12	14	35	50	24
11	04	22	63	43	66	97	86	12	65	36	85	17	19	62	38	22	19
41	88	11	92	84	89	64	14	15	67	49	59	07	06	54	62	63	10
62	03	31	01	38	33	52	46	16	75	57	97	46	53	38	29	01	48
41	60	86	40	45	12	19	75	86	37	57	11	08	45	12	66	70	92
06	14	52	22	63	13	54	70	46	48	06	01	39	55	04	92	37	72

Random Number Table 2

33	42	15	26	04	13	23	06	65	25	94	72	07	52	25	16	33	09
75	83	32	13	78	02	02	62	21	98	99	98	28	58	08	26	60	83
08	71	32	05	57	43	11	97	87	34	05	56	46	90	47	36	44	24
94	22	72	68	51	54	08	48	69	15	43	34	72	56	04	51	42	30
40	04	38	81	52	15	06	66	13	27	67	22	43	21	38	90	98	97
51	96	61	81	95	44	66	30	88	32	12	50	30	23	02	35	20	27
09	61	25	34	95	83	76	57	84	47	77	96	70	80	46	71	74	07
73	78	56	55	69	72	40	17	14	94	96	93	58	01	38	51	64	48
82	92	49	41	17	92	54	59	26	82	64	13	25	99	37	41	78	20
54	53	18	77	22	27	18	20	36	79	26	54	52	78	77	92	94	39
04	23	70	21	40	56	53	31	28	63	07	56	11	89	80	78	62	28
11	29	18	50	89	82	13	37	97	28	11	72	78	81	35	28	69	52
71	56	10	97	31	80	36	27	95	54	87	58	59	01	38	25	65	78
83	86	82	30	15	36	66	14	47	86	61	37	41	60	39	44	77	45
64	86	75	07	25	86	41	67	18	73	38	10	12	51	78	29	32	23
44	07	49	46	80	71	01	76	25	63	37	45	05	54	90	68	55	33
67	24	47	55	03	34	68	09	89	65	99	67	81	46	33	13	20	19
97	39	51	11	69	66	27	51	59	95	23	03	97	24	56	45	50	49
46	52	77	86	68	59	55	42	14	20	79	65	26	42	14	81	77	05
80	88	31	64	30	76	65	27	59	24	61	43	18	34	19	16	90	66
82	32	25	55	11	03	68	07	75	78	49	86	43	02	80	09	21	25
65	43	63	54	92	10	64	14	18	60	78	23	49	12	78	45	07	66
37	15	07	30	79	55	28	30	22	83	30	48	14	32	47	71	17	39
04	09	40	13	55	74	93	68	34	46	66	97	40	40	59	54	72	62
55	01	96	04	74	89	44	05	06	27	35	45	87	23	60	45	33	03
34	34	78	45	30	34	66	96	97	03	67	29	72	55	85	28	45	46
30	99	83	22	22	78	05	18	10	31	49	76	82	70	01	57	58	52
99	32	49	58	85	25	13	17	48	37	69	88	97	12	63	88	31	17

Contd...

Random Number Table 2

98	43	81	48	54	43	21	54	79	45	29	42	13	84	18	81	79	31
30	01	71	27	61	78	07	86	82	19	97	95	27	65	48	82	92	72
57	11	30	34	51	28	22	36	15	75	16	93	31	97	52	91	77	75
90	75	29	61	06	27	83	05	53	92	57	21	12	53	83	35	05	20
56	03	19	18	06	34	29	71	31	43	27	93	30	72	66	73	35	64
38	15	39	20	66	20	62	23	85	92	65	82	83	71	04	13	71	52
39	67	68	68	86	58	76	16	96	74	16	66	41	77	20	44	66	88
47	83	50	92	70	84	55	29	99	69	86	55	06	73	35	35	23	80
91	65	38	41	58	55	87	10	24	01	04	64	06	16	79	15	48	35
69	32	95	46	04	35	13	45	17	01	08	09	55	79	17	65	82	44
80	05	97	80	71	44	59	37	44	38	51	81	25	97	35	63	26	88
12	36	51	05	62	27	34	60	99	43	12	22	17	50	89	32	82	17

APPENDIX A

Location and Address of Regional Centres of National Afforestation and Eco-development Board

The regional centres of the National Afforestation and Eco-development Board (NAEB) have been created as decentralized organs of the NAEB to facilitate training, research, and outreach for different states. A list of the centres is as follows:

1. Regional Centre for NAEB, University of Agricultural Sciences, GKVK Campus, Bengaluru-560065

   Communication Linkages—Tel: 080-23334210; Fax: 080-23330277; email: rcnaeb@indiatimes.com

   Area: Andhra Pradesh, Karnataka, Kerala, Lakshadweep, Puducherry, and Tamil Nadu
2. Regional Centre for NAEB, Dr Y. S. Parmar University of Horticulture and Forestry, College of Forestry, Nauni, Solan-173230

   Communication Linkages—Tel: 01792-252487; Fax: 01792-252242

   Area: Chandigarh, Himachal Pradesh, Jammu and Kashmir, and Punjab
3. Regional Centre for NAEB, Agricultural Finance Corporation Ltd, B-1/9, Community Centre, Janakpuri, New Delhi-110058

   Communication Linkages—Tel: 011-25550810; Fax: 011-25597437; email: afcdel@nde.vsnl.net.com

   Area: Delhi, Haryana, Rajasthan, Uttar Pradesh, and Uttarakhand

4. Regional Centre for NAEB, Indian Institute of Forest Management, Nehru Nagar, Post Box No. 357, Bhopal-462003

   Communication Linkages—Tel: 0755-2775716; Fax: 0755-2772878

   Area: Chhattisgarh, Madhya Pradesh, and Odisha

5. Regional Centre for NAEB, Agricultural Finance Corporation Ltd, Dhanraj Mahal, 1st Floor, CSM Marg, Mumbai-400001

   Communication Linkages—Tel: 022-2028924; Fax: 022-2028966; email: afcl@bom2.vsnl.net.in

   Area: Daman and Diu, Dadra and Nagar Haveli, Goa, Gujarat, and Maharashtra

6. Regional Centre for NAEB, North-Eastern Hill University, Shillong-793014

   Communication Linkages—Tel: 0364-2550979; Fax: 0364-2551599

   Area: Assam, Arunachal Pradesh, Manipur, Meghalaya, Mizoram, Nagaland, and Tripura

7. Regional Centre for NAEB, Jadavpur University, Kolkata-700032

   Communication Linkages—Tel: 033-24146979; Fax: 033-24146886; email: rcnaeb@vsnl.net

   Area: Andaman and Nicobar Islands, Bihar, Jharkhand, Sikkim, West Bengal

APPENDIX B

National Afforestation Programme at a Glance

BACKGROUND

The Ministry of Environment and Forests (MoEF) was operating four centrally sponsored afforestation schemes during the Ninth Five-year Plan: (1) Integrated Afforestation and Eco-development Project, (2) Area Oriented Fuelwood and Fodder Project, (3) Conservation and Development of Non-timber Forest Produce including Medicinal Plants Scheme, and (4) Association of Scheduled Tribes and Rural Poor in Regeneration of Degraded Forests. The mid-term evaluation of the projects under these schemes recommended a decentralized approach and speedy fund transfer mechanism. Taking cue from these recommendations, a pilot scheme, titled Samanvit Gram Vanikaran Samriddhi Yojana (SGVSY), was launched in 2000/01 utilizing the delivery mechanism of the forest development agencies (FDAs) and joint forest management committees (JFMCs). Forty-seven pilot projects launched during the Ninth Five-year Plan were highly successful in terms of achievement of the set objectives.

According to the Tenth Five-year Plan document of the Planning Commission relating to the forests and environment sector (para 9.1.27), the National Afforestation Programme (NAP) was initiated by scaling-up the SGVSY project experience and converging all afforestation schemes of the Ninth Five-year Plan period to avoid duplicity or redundancy and, at the same time, keeping in focus the decentralization agenda of the government. NAP is being operated as a 100% central sector scheme.

Some FDAs started afforestation in 2000/01, and it was envisaged that around 0.7–0.8 million ha area will be brought under plantation by the end of the Tenth Five-year Plan through FDAs.

OBJECTIVES

The overall objective of the scheme is to develop forest resources with people's participation and with a focus on improvement in livelihoods of communities living in the fringe of forests, especially the poor.

The NAP aims to support and accelerate the ongoing process of devolving forest protection, management, and development functions to decentralized institutions of the JFMCs at the village level and the FDAs at the forest division level.

COMPONENTS

Financial support under the NAP scheme is available for the following purposes.

- Mobilization of village JFMC and microplanning in project villages.
- Afforestation with the following components.
  - Aided natural regeneration.
  - Artificial regeneration.
  - Bamboo plantation.
  - Cane plantation.
  - Mixed plantation of trees having minor forest produce and medicinal value.
  - Regeneration of perennial herbs and shrubs of medicine value.
  - Pasture development/silvopasture.
- Soil and moisture conservation.
- Entry-point activity (for village development; average assistance of ₹0.16 million per village)[1]
- Fencing, monitoring and evaluation, training, awareness raising, overheads.

INSTITUTIONAL FRAMEWORK AND ITS STRENGTHENING

The NAP is being implemented through a two-tier structure of FDAs at the forest division level and JFMCs at the village level. Thus an FDA is the confederation of JFMCs in that forest division. FDAs are registered under the Societies Registration Act. JFMCs are registered either with the Forest

[1] For updated statistics, please visit the MoEF website <www.envfor.nic.in>

Department or under statutory provisions. The district-level officers of relevant line departments of the state government and *Panchayati Raj* institutions are members of the FDAs.

The recent experience with the present institutional frameworks of FDAs and JFMCs is encouraging. All 28 states have adopted this institutional framework, including the autonomous hill districts of Assam, Meghalaya, and Mizoram.

The institutions of FDAs and JFMCs are highly innovative resource transfer mechanisms whereby the Government of India channelizes funds directly to the grass-root level implementing agency for afforestation activities.

The structures of FDAs and JFMCs also cater to gender concerns, whereby women membership to the extent of 50% has been made mandatory. Members of schedule castes and schedule tribes are the focus groups in JFMCs.

The training of FDA and JFMC members is organized by state forest departments, as well as by the seven regional centres of the National Afforestation and Eco-development Board (NAEB).

To help and guide the FDAs and JFMCs, there is also a national-level steering committee of the NAP and another at the state level chaired, respectively, by the secretary of environment and forests, Government of India, and the chief secretary of the state government.

The following additional steps are being undertaken to further strengthen these institutions.

- An assessment exercise for training needs is being undertaken by the regional centres of the NAEB to identify the needs and capacities of different categories of staff and members of JFMCs and FDAs for efficient implementation of NAP, and subsequently to develop training modules.
- It has been proposed to encourage linkage of JFMCs with other rural development programmes. JFMCs may leverage outside funds to augment the entry-point activity component of NAP for this purpose.
- It has also been proposed to formulate a financial management manual for FDAs and JFMCs so as to strengthen account keeping at the local level and thus to empower communities for local-level decision-making.
- A national-level joint forest management (JFM) study has also been launched to assess the strengths and weaknesses of implementation of JFM and the progress of empowerment of the local communities in decision-making.
- A system of e-governance in the NAEB has been established.

MONITORING AND EVALUATION MECHANISM

- With a view of increasing the efficiency of NAP to meet the aforementioned goals, the monitoring and evaluation (M&E) exercise of the FDA projects is undertaken both by the state government and by the central government.
- The NAEB undertakes the first independent mid-term evaluation of FDA projects within 24 months of sanction of the project with a focus on people's participation, functioning of JFMCs/eco-development committees, and the microplanning exercise.
- The final evaluation is required to be carried out in the fourth year of the project.
- So far, the NAEB has launched mid-term evaluation in 250 FDA projects (including 47 projects under the SGVSY of the Ninth Five-year Plan).
- In addition, the Forest Survey of India (FSI) randomly selected 60 FDAs each in 2003/04, 2004/05, and 2005/06 to monitor area coverage and forest cover increase therein.

Monitoring of Area Coverage and Survival Percentages of Plantation/afforestation

The FSI, Dehradun, prepared a four-year project to monitor the plantations under the NAP, commencing from 2003 and culminating in 2007. Sixty FDAs selected randomly across the country were monitored every year. It was envisaged that by the end of 2007, 240 FDAs, comprising around 1000 patches, would be monitored. The main objectives of the project are as follows.

- Monitoring of the area coverage and estimation, at a national level, of the area coverage of selected FDAs.
- Estimation, at the national level, of the survival percentage of plantations (species wise) raised by FDAs inside and outside forest areas.
- Assessment of the growth parameters of trees planted under the NAP.

Methodology

The FDA areas of the districts were taken as the study area for this survey. Stratified random sampling was employed and FDAs were selected randomly from the whole country. The patches of FDAs were further stratified into five strata—four strata in forest area based on patch area (0–25 ha, 25–50 ha, 50–75 ha, and 75+ ha) and one stratum in non-forest area, which were the sampling units. On the basis of the pilot study, the sample size for monitoring FDAs in plain and hilly areas was determined.

The selected patches within an FDA were physically verified with the help of Global Positioning System by traversing the whole boundary. The expected results of the survey will be with 15% permissible error at 95% probability level at the national level. The following data were collected from the selected on FDAs.

- Area of patches.
- Total number of plants planted in the plot.
- Total number of plants survived in the plot.
- Name of tree species, height, and number of diseased/dead plants in the plot.

The survival percentage was calculated from the data collected from the laid out plot of 0.1 ha in the patches for both hilly and plain areas, that is, the total number of plants planted and the number of plants that survived in the patch. During 2003/04 and 2004/05, 60 FDAs each were chosen for monitoring. The analysed results of these districts had been sent to the MoEF. During 2003/04, the area coverage under FDAs at the national level was estimated to be 93.29% with 3.79% standard error (SE).

Similarly, the survival percentage of plantation was estimated to be 87.58% at the national level with 5.33% SE. The results of 2004/05 indicated that the area coverage at the national level was estimated to be 97.34% with 2.17% SE and the survival percentage was estimated to be 87.23% with 5.02% SE.

ACTIVITIES UNDERTAKEN SO FAR

As on 31 March 2006, 715 FDAs were made operational so far at a cost of ₹15,147.9 million to treat a total area of 0.923 million ha. Bamboo plantation, medicinal plants, and Jatropha were given adequate focus under NAP during the current plan period. State governments were advised to earmark 10% of the project area under NAP, according to feasibility, for Jatropha plantation. Rehabilitation of *jhum* lands (shifting cultivation) was given specific focus under the programme, and so far 19 *jhum* projects had been sanctioned in north-eastern states and in Odisha.

PROGRESS/ACHIEVEMENT MADE DURING 2005/06

Ninety-five new FDA projects were sanctioned to cover an area of 55,232 ha through 2391 JFMCs. These include 12 new FDA projects in the north-eastern states to cover a total project area of 5380 ha through 183 JFMCs. An amount of ₹2485.8 million was released to FDAs during 2006, out of which the release to north-eastern states was ₹458 million (up to 31 March 2006).

PHYSICAL OUTPUTS AND EXPECTED OUTCOMES

The goal of NAP was to promote improvement or increase in forest and tree cover. Two outcome parameters were, therefore, identified with respect to NAP.

1. Five years after sanction, new plantations would, for each biogeographic region, show the prescribed survival rates.
2. Six to twelve years after sanction, depending on the species, the new plantations would show as either new area under forest and tree cover or enrichment of forest cover in satellite imagery.

COMPARISON OF PROGRESS AS COMPARED TO PREVIOUS YEARS

Year-wise progress of the NAP in the Tenth Five-year Plan is given in Table A. Table B shows the progesses of FDA projects (up to 31 March 2006).

Table A Progress of the NAP in the Tenth Five-year Plan

Year	*Number of new FDA projects operationalized*	*Number of JFMCs*	*Project area (ha)*	*Release (₹ million)*
2002/03	237	8,209	405,631	1,512.6
2003/04	231	7,850	283,272	2,079.8
2004/05	105	3,474	107,963	2,330.0
2005/06				
(up to 31 March 2006)	95	2,391	55,232	2,485.8

Table B Progress of FDA projects from April 2000 onwards (as on 31 March 2006)

Name of state/ union territory	*Number of FDA projects/proposals sanctioned*	*Total project cost (₹ million)*	*Area (ha)*
Andhra Pradesh	32	830.2	47,400
Chhattisgarh	32	738.3	42,514
Gujarat	21	608.7	30,445
Haryana	18	534.4	22,105

Contd...

Table B *Contd...*

Name of state/ union territory	*Number of FDA projects/proposals sanctioned*	*Total project cost (₹ million)*	*Area (ha)*
Himachal Pradesh	27	523.8	32,378
Jammu and Kashmir	31	716.1	17,839
Karnataka	45	1,121.5	59,180
Madhya Pradesh	49	1,124.8	76,520
Maharashtra	45	986.2	65,738
Odisha	40	696.5	55,022
Punjab	9	176.5	8,235
Rajasthan	33	392.2	28,190
Tamil Nadu	32	936.0	52,253
Uttar Pradesh	58	1,043.5	63,004
Uttaranchal	37	541.7	39,134
Goa	3	23.9	1,250
Jharkhand	30	563.3	39,850
Bihar	10	161.2	12,315
Kerala	24	475.4	16,250
West Bengal	17	390.7	22,856
Total (other states)	593	12,614.9	762,478
Arunachal Pradesh	19	270.4	21,416
Assam	29	365.6	26,955
Manipur	13	265.8	18,374
Nagaland	16	377.1	25,528
Sikkim	7	277.2	15,280
Tripura	12	255.7	19,405
Mizoram	19	601.2	26,770
Meghalaya	7	120.0	7,400
Total (north-eastern states)	122	2,533.0	161,128
Grand total	715	15,147.9	923,606

BUDGET ALLOCATION OF THE SCHEME AND PROGRESS OF EXPENDITURE

Revised estimate for the scheme during 2005/06 was ₹2485.8 million, which was released till 31 March 2006.

IMPLEMENTING ORGANIZATION ALONG WITH DETAILS

The NAP is being implemented through a two-tier decentralized mechanism of FDAs at the forest division level and JFMCs at the village level.

NEW INITIATIVES UNDER THE SCHEME

A number of initiatives were taken by the ministry to expedite the implementation of the scheme as well to improve the qualitative aspects of implementation. These initiatives include the following.

- A decision was made by the national-level steering committee for the electronic transfer of funds from the Government of India to the FDAs to cut down delays.
- Efforts were made to step up the M&E of the FDA projects by the activation of state-level coordination committees for NAP, increased field visits by officers, and expeditious commissioning by the states of the first independent concurrent evaluation of FDA projects.
- Ten programmes were arranged through the Directorate of Forest Education for training and capacity building of front-line staff of FDAs and JFMCs.
- Forty-five district-level interdepartmental linkage workshops were approved in the work programme of the regional centres of the NAEB for 2005/06 and 2006/07 to promote the linkage of NAP with other developmental programmes for enhancing the sustainability of JFM.
- A new component was added in NAP to universalize JFM in all forest-fringe villages in the country.

MONITORING AND EVALUATION OF NAP

It has been proposed to undertake evaluations of the ongoing projects and schemes of NAEB to assess achievement against the set objectives as well as to evaluate the strengths and weaknesses of various technologies used so that appropriate amendment in the current schemes of the NAEB can be made for greater efficiency. It has also been proposed to continue with the following.

- The first concurrent evaluation of FDA projects under the NAP and projects under the Grants-in-Aid to Voluntary Agencies scheme.
- Study of tree planting under the Twenty-Point Programme in selected districts of India.

- Monitoring of area coverage and survival percentage in selected FDAs by the FSI.
- Interactive meetings on the M&E of FDA projects under NAP at the regional level.
- Mid-term evaluation of the schemes operated by the NAEB to assess their feasibility for continuation in the Ninth Five-year Plan.

A sum of ₹17.5 million was proposed for 2006/07 for the rest of India. The M&E of schemes/projects in the north-eastern states would also be undertaken with the proposed sum.

Monitoring of Area Coverage and Survival Percentages of Plantation/afforestation under the NAP at Macro Level

The NAEB entrusted the work of monitoring the progress of area covered by plantations and the survival percentage under the NAP by these FDAs on forest and non-forest lands to FSI. Accordingly, FSI prepared a project for four years as explained before.

Source FSI (Forest Survey of India). 2006. An overview on projects. In *Forest Survey of India, 2006*. Dehradun: FSI

Index

About the Editors and Contributors

V. K. Sinha has worked as a senior professor in ecosystem management and technical forestry at the Indian Institute of Forest Management (IIFM), Bhopal. He has extensively worked in various key positions as forest manger in the state of Maharashtra and the Government of India, including regional manager of the Maharashtra Forest Development Corporation at Nagpur. He is a practising forester, principally working on forest policy and planning issues. He is currently working as Field Director of Tadoba Tiger Reserve, Maharashtra.

H. S. Gupta is currently professor in ecosystem management and technical forestry at IIFM, Bhopal, with current research focus on collaborative working in forestry projects, plantation forestry, and management wildlife. He is an Indian Forest Service (IFS) officer of Jharkhand cadre, 1986 batch, and has worked extensively in different districts of Bihar and Jharkhand, carrying out various models of plantations/afforestation, and has analysed the different dimensions of monitoring and evaluation. His other interests include sustainable forest management (SFM), certification, and climate change related to forestry.

R. K. Singh is currently Director of the Chhattisgarh Institute of Rural Development at Raipur, Chhattisgarh. He has worked as a senior professor in ecosystem management and technical forestry at the IIFM, Bhopal. He is a senior practising forester whose research interest is in capacity building, rural development, and collaborative techniques aspects of forestry and natural resource management.

D. K. Sharma belongs to the 1988 batch of the IFS Manipur–Tripura cadre. He is Deputy Inspector General (forests) in the Ministry of Environment and Forests, Government of India, and has served in the past as chief conservator of forests and director, NTFP Centre of Excellence, Tripura. Dr Sharma has written many articles and books on PA governance, environmental issues, management of non-timber forest products, forest fire issues, and improvement in forestry sector.

S. K. Barik is currently a senior faculty at the North-Eastern Hill University, Shillong. His research and teaching activities include ecology and natural resource management.

J. Kulkarni is associated with Envirosearch, Pune, as a consultant for forestry, environment, and other related assignments. His research interest includes evaluation and monitoring of forestry projects, besides wildlife and ecology-related issues.

N. Sahu is a monitoring and evaluation expert. He has extensive experience in watershed-related projects, particularly in the state of Odisha. He has contributed in the present book as Monitoring and Evaluation Specialist, Project Support Unit, Western Orissa Rural Livelihoods Project, Bhubaneswar.

G. B. Reddy is a senior forester with long experience in afforestation and plantation work in the state of Odisha. He also has the inside experience of using watershed activities for livelihood promotion of the poorest of poor villagers by directing one such project very successfully, namely the Orissa Watershed Development Mission. He has contributed in this book as Director, Orissa Watershed Development Mission, Bhubaneswar.

C. V. R. S. Vijaya Kumar is currently teaching quantitative techniques as associate professor in IIFM, Bhopal. He got his PhD in statistics (specialization in queuing theory) in 1998 and MSc (statistics) in 1991, both from Andhra University. His areas of academic interest are operations research and stochastic modelling, with particular emphasis on forestry sector.